THE WORLD OF MAPS

Also from Judith A. Tyner

Principles of Map Design

THE WORLD
OF MAPS

Map Reading and Interpretation
for the 21st Century

Judith A. Tyner

THE GUILFORD PRESS
New York   London

© 2015 The Guilford Press
A Division of Guilford Publications, Inc.
72 Spring Street, New York, NY 10012
www.guilford.com

Printed in the United States of America

This book is printed on acid-free paper.

Last digit is print number: 9   8   7   6   5   4   3   2   1

Library of Congress Cataloging-in-Publication Data is available from the
publisher.

ISBN 978-1-4625-1648-3 (cloth)

To Gerald Tyner,
for many years of moral support
and patience

Preface

I became interested in maps at age 7 when I asked, "How much farther?" one time too many. My mother handed me the road map, showed me how to read the basics, and said, "You figure it out." I've been in love with maps ever since.

Of course, there is more to map reading than figuring distances, there are more kinds of maps than road maps, and there have been many changes in mapping and cartography since that long ago road trip. It is my goal to create a user-friendly book that can be used in classes, but can also used by the interested layperson and kept handy on the shelf for a quick reference. I hope to introduce readers to the intriguing world of maps and let them see just how much maps can tell us. To achieve that goal, the topics are dealt with in stand-alone "bites." There is more information in this book than would normally be covered in a semester or quarter, but this allows instructors to insert some topics of special interest to them and their students.

One might ask, "Is a book on map reading necessary in the digital age when virtual maps are available?" We have mapping sites on the web that will plot a route for us, and we have hand-held and in-car global positioning systems (GPS) that will plot routes, so why do we need any other maps? First, virtual maps and GPS maps are still maps and require knowledge of scale, coordinates, and the like. Second, like all map forms, these have limitations, and one needs to understand those limitations or risk getting lost, or worse. Finally, there are still hundreds of different map types that one encounters on a daily basis in newspapers, magazines, textbooks, and advertisements. To rephrase the statement above, reading road maps is not the only way of using maps.

I have divided the book into three major sections. Part I, Map Reading Basics, includes an introduction to maps and map use; the core concepts or basic skills of latitude and longitude, projections, scale, and direction; kinds of maps; and a brief history of cartography. Part II, Map Types and Their Analysis, introduces the reader to different map types such as topographic maps, navigation maps, and thematic maps, and examines how to read and analyze each type. Part III, Putting It All Together, explains how to use multiple maps and imagery to analyze an area or region. There is some redundancy built into the book, with the same fundamental topics sometimes approached from different perspectives in different chapters. For example, Chapter 3, Map Basics, includes information about various symbol types. These symbols are also discussed in Chapter 8, Topographic Maps, and Chapter 9, Thematic Maps, as they relate to those map types.

A note on units of measurement: English (Imperial) units such as miles and yards are used primarily, and metric units such as kilometers and meters are given in parentheses. However, when metric measurements are not appropriate—for example, the dimensions of a baseball diamond or, more importantly, the dimensions of the Public Land Survey System (Chapter 4), which are specifically in miles and acres—I have included only English units.

Several appendices at the end of the book present resources and useful tables.

A NOTE ON THE CHAPTER-OPENING EPIGRAPHS

Each chapter opens with a quotation from either David Greenhood's *Mapping* or *Down to Earth*.[1] Why are all of the epigraphs from a single author? Who was David Greenhood? In 1944, in the thick of World War II, Greenhood, a lifelong lover of maps and a self-proclaimed "map muser," wrote a book for the layperson, the amateur user and maker of maps. It was a time when maps were important to the public, who followed the course of the war in newspapers and magazines. Twenty years later, *Down to Earth* was followed by a new edition, *Mapping*, with the assistance of Gerard L. Alexander of the New York Public Library, but the book retained Greenhood's "voice." *Mapping* is still in print 50 years later, and unless otherwise noted, the quotations are from that edition. Much of the material in the two editions is still relevant and useful, despite being written before the advent of personal computers and GPS. Many statements in the books anticipate current concerns and technologies. For example, Greenhood pointed out map limitations and the use of maps for propaganda. "People are plain suckers if they never question the reliability of maps shoved under their noses. The result might be the total loss of civil liberties . . ." (1944, p. 5).

In *Down to Earth*, Greenhood even anticipated what it would be like to see the earth from the moon.

> As we stand on the moon, looking at the earth, we see a colorful shining globe against a black sky. It is a heavenly body of great beauty. Shimmering oceans with white clouds afloat. Sparkling polar caps. It looks so lively, so full of animation that it seems to call us. (p. 1)

---

[1] Quotes from David Greenhood, *Mapping*, copyright 1964 by the University of Chicago, are used with the kind permission of the University of Chicago Press.

We picture the early photographs from the moon in our mind's eye and then remember with a jolt that when Greenhood wrote that paragraph the moon landing was over 25 years in the future (see Plate 6.2).

These were the only mapping books written by Greenhood. He also wrote novels, some children's books, and, especially, poetry. He didn't work for a mapping agency or company; he and his wife founded a publishing company, Holiday House, that specialized in children's books and also published *Down to Earth*.

In this work I pay homage to David Greenhood, from whom I've learned that perhaps it takes a poet to remind us to take delight in maps and "make ourselves more at home in the universe."

Acknowledgments

Although this book has only one author's name on it, it could not have been written without the assistance of people with a wide range of interests and knowledge.

There were many who helped during the process: I thank Greg Armento, Map and Geography Librarian at California State University, Long Beach (CSULB), for finding maps and references, and my colleagues at the CSULB Geography Department for their willingness to answer off-the-wall questions on short notice. I especially thank Norman Thrower and Mark Monmonier, from whom I've learned so much through chats by phone, over coffee, and by e-mail. Map lover Steve Spangler doesn't realize that our casual chats about maps were actually helping me organize the material. Of course, I must acknowledge the late Richard Dahlberg, from whom I took my first map reading and cartography classes, and the late Gerard Foster, who mentored me in teaching map reading.

I want to thank my family, who put up with me during the process: James Tyner for many discussions about writing and David Tyner for his unfailing encouragement of my work. Gerald Tyner deserves special thanks not only for moral support (and critiquing) during the writing, but also for creating the majority of the maps and diagrams.

Finally, I want to recognize the team at The Guilford Press for their patience and encouragement: Kristal Hawkins, who commissioned the project; C. Deborah Laughton, the final editor; and staff members Mary Beth Anderson and William Meyer.

While all of these people helped make this book possible, any flaws or errors are my own.

Contents

Part I. Map Reading Basics

Part II. Map Types and Their Analysis

Part III. Putting It All Together

Downloadable PowerPoint slides of selected figures are
available at *www.guilford.com/tyner3-materials*.

PART I

Map Reading Basics

CHAPTER 1

Introduction: The Importance of Map Reading

> There are often many truths in a place or an area
> right before our eyes, and yet we're not aware of
> those truths (or features, or facts) until a depiction
> or a symbol or even a diagram *shows* them to us.
>
> —*Mapping* (pp. x–xi)

We see maps every day, often without really looking at them. A weather map is shown on a morning TV news show, or a map pinpoints the location of a natural disaster, rush hour traffic is displayed with colors relating speed and symbols indicating "traffic incidents." The morning paper, in addition to a weather map, might have a U.S. map that shows unemployment by state or worldwide incidence of the latest flu strain. All of these maps have been presented to us before we finish our morning coffee. As the day progresses, we might use an online map to find our afternoon appointment, an in-car GPS map for turn-by-turn directions, or a street atlas if we don't have access to GPS. A transit map might show us which train to take and the relation of our exit station to other stations. At work we might need a map to help decide where to put a new store or where to allocate money. And so it goes. Throughout the day maps pass before our eyes. Maps are ubiquitous.

But do we really get the most from these various maps? Do we stop to read them carefully? Other than knowing what the weather in our town will be today, whether the natural disaster will affect us, or if the traffic incident is on the freeway we normally drive, what have we learned from these maps? Could we use them more effectively?

Given the commonness of maps and their many uses, not being able to read maps effectively is like only being able to read text at a third-grade level. We might be able to understand the basics, but much is lost to us.

First, although we all "know" what a map is, it is helpful to define the term as cartographers use it. Cartographers are mapmakers, and cartography is the art, science, and technology of making maps and also their study as historical documents and/or works of art. But what *is* a map? J. H. Andrews discovered 321 definitions of "map" for his article "What Was a Map?" (1999). The most common definition is: A graphic representation of all or a part of the earth or other body, drawn to scale upon a plane. However, this definition limits us because some objects we recognize as maps, such as sketch maps, are not drawn to scale, nor are some maps of preliterate peoples. Some "maps" are annotated photos or imagery, not drawn graphics. Some maps on a computer monitor or cell phone are animated. Thus, for our purposes we will define maps as spatial representations of information.

Second, we should consider the purpose or goals of maps. Maps *represent*; that is, they portray information about a place symbolically. Maps also are a form of *communication*: The cartographer conveys information to a map user. The map is the medium of communication. Maps are *visualizations*; visualization refers to exploring data and seeing data in different ways. While a series of maps of the same data could be considered visualizations, usually the term is associated with dynamic visual displays on a computer screen; the goal of visualization is to gain insight into the data. Finally, maps are *arguments*, as suggested by Denis Wood (2010, pp. 42–44); maps argue their points and can be thought of as rhetorical devices.

CHANGES IN THE PAST 25 YEARS

There is a tendency to think that maps are maps and "if you can read one, you can read 'em all." In reality, enormous changes have taken place in maps and mapping in the past 25 years. We owe many of these changes to the computer; some map types that are now common, such as animated maps, were rarely seen because of the difficulty of drawing them and viewing them. The computer has changed the methods of creating maps, of viewing maps, of delivering maps, and of publishing maps. Thirty years ago if you wanted to go for a hike, you might have purchased a topographic map—one showing the nature of the land, hills, and valleys—from an outdoor recreation company, or you might have ordered it from the U.S. Geological Survey (USGS). Now you can download a topographic map from the USGS, or you can buy topographic software from companies such as Delorme or National Geographic that will help you plan your hike and even provide profiles of your route. You might even download large-scale topographic maps to your hand-held GPS (global positioning system) unit and navigate and create profiles of your trail with an apparatus that fits in your pocket.

Maps of many types are on the Internet; some of these maps are interactive, and you can query the map to gain information not available on a static map. Others are animated and show changes through time; haptic maps allow you to "feel" textures, and sound maps speak information to the user.

The changes in maps and mapmaking are more in their creation and delivery than in techniques of reading and using. We can download a book to an e-reader, but we still read the words in the same manner as we read words on paper. The computer and geographic information systems (GIS) have revolutionized the ways in which maps are made, and the computer and the Internet have introduced new ways of delivering maps to users. All the same, basic map reading techniques can still be used for most map reading tasks, although some "new" map types—animated, sound, and haptic—do require looking at maps in new ways and using different methods.

EVALUATING MAPS

Not all maps are created equal; as we shall see below, there are many hindrances to using a map. One of these obstacles is the quality of the map itself. A poorly drawn map can sometimes tell its story effectively, but an elegantly drawn map created from poor data may be worse than useless: It can mislead. I hope that the reader will be an informed and discriminating map user after reading this text.

MAP LIMITATIONS

All maps have limitations. This is equally true for a hand-drawn map of the 12th century and for the most up-to-date computer-created map. It is true for paper maps and for electronic virtual maps. It is true for paper road maps and for maps on a GPS. The limitations are not always obvious or deliberate, nor are they necessarily bad. They come about in a variety of ways and stages in the mapmaking process—the nature of maps and of the data, the tools used, the skills of the mapmaker and map user, and the biases of the mapmaker and agency or organization making the map. In this section we look at limitations in general; limitations of specific map types are discussed in each section (see Table 1.1).

Limitations from the Nature of Maps

Maps are usually graphic, not photographs; they are usually drawn to scale, symbolic, and flat. Each of these factors creates limitations. Photographs show everything within the view of the lens; maps are drawings and they are selective. This can be a limitation in that the map isn't inclusive. Maps are drawn to *scale*, which means that a large area is reduced in size. On a world map, the 3,794,083 square miles (9,826,630 square kilometers) of the United States might be reduced to a square inch (or square centimeter). It is not possible to show everything on any map; therefore, maps are *generalized*. The amount and type of information to be shown are selected, data are put into categories, and complex features are smoothed and simplified. Maps are *symbolic*; that is, they use icons, colors, shades, and lines to represent information. Maps are flat and the earth is round. The spherical earth is *projected*, that is, converted in a systematic and orderly way to a flat surface. Since a sphere cannot be flattened without stretching, tearing, or shrinking, flat maps have *distortions* in area, shape, distance, or angles.

TABLE 1.1. Map Limitations

The nature of maps	Cartographer limitations
Scale	Skills
Projection	Knowledge of subject
Generalization	Biases
Data limitations	Agency or client limitations
Data accuracy	Objectives
Positional accuracy	Biases
Lack of data	
Gaps in data	Map user limitations
Currency of data	Lack of skills
	Stereotypes
Technology limitations	Using wrong map for purpose
Drawing/creation methods	
Delivery method	
Printing	
Hardware	
Software	

Limitations Introduced by Data

Maps are made from data. These data may be positional, involving the latitude and longitude of a place, an address, a road, or a border. Some maps use statistical data to show such things as population; others use qualitative data such as locations of grasslands or deserts. Errors can enter when data are collected. Maps made by geographic information systems (GIS), maps on the Internet, and GPS units must be *digitized*. That is, the data must be put in a form that the computer can read. Some of this digitizing is done by humans plotting and entering the information, and thus, there is the potential for human error. On historic maps and even some modern maps there might be lack of data, gaps in the data, and even erroneous data.

One must also remember that paper maps and even electronic maps are made at a specific time and that a printed map could be many years old. Therefore, the information on the map is valid only for the date of production. Even electronic maps, GPS base maps, and maps on the web may not have the most up-to-date information; they are also limited by the date of production.

Limitations Introduced by Technology

Through time, maps have been created using many different kinds of tools and different materials. A map's appearance and content are often determined by the tools the mapmaker uses. The oldest maps in existence were carved into clay tablets with a pointed tool, and others were incised on rock. Maps have been made on papyrus, cloth, metal, sheepskin, paper, and plastic. And, of course, maps are "made" on a computer screen.

Beginning in the second century C.E., maps were drawn on paper using pens, ink, and drawing tools such as rulers and compasses. Before the invention of printing, maps were copied over and over by hand; inevitably, errors crept into the copies, and no two were exactly alike. With the advent of printing, maps were exactly reproducible, but different methods of printing—woodcut, copper plate, lithography, and offset lithography—introduced new limitations. Woodcut maps tended to be less detailed than copper plate maps; color was difficult to use with these two methods and was applied by hand. Lithography allowed shading, and color was relatively easy to print. Offset lithography permitted rapid production of thousands of copies.

In the last third of the 20th century, computers were introduced into mapmaking, and today the majority of maps are made using computer *software*. Again, there are limitations. Software used for mapping is of several types: dedicated mapping software, GIS, and illustration or drafting software. Each of these has limitations. GIS is a powerful analytical tool for spatial data and is widely used for mapmaking. "Smart maps" can be made using GIS: Such a map has information associated with each point, line, or area on it, and the map can be "queried" to find that information. But some kinds of maps can't be made with GIS, and the map is only as good as the software capabilities and the user's skills. If no analysis of data is required, the cartographer may use illustration software such as Adobe Illustrator or CorelDraw, or he/she may combine two types of software. As a reader, you do not always know how a map was created.

Limitations Introduced by the Mapmaker

The mapmaker's skills with drawing equipment or software has a bearing on the final map. Did the map drafter draw an angle slightly off, was the GIS technician knowledgeable about cartography and maps, or was he/she following default options of the software? Maps are not always created by professionals who have knowledge of cartography or of GIS systems. The mapmaker might not have any knowledge of the subject being mapped but may simply be entering data into a software program. Mapmakers have biases. In the words of John K. Wright (1942), "mapmakers are human" (p. 527).

Limitations Introduced by the Map Agency or Client

Maps are not neutral. Indeed, often maps are created by people with an "axe to grind." An organization might want to stop a commercial development and will create maps that emphasize its point of view; a transportation company might want to show that its routes are shorter; while another company might want to show that it is more centrally located than a competitor. In all of these cases, some information might be left out or deemphasized, and other information might be distorted or enhanced. In the early days of railroads, for example, companies created distorted maps that straightened out their routes and changed the sizes of states to "show" that their routes were more direct and quicker. Maps are often made for advertising, and the company wants to emphasize its products. These maps are examples of *persuasive cartography* in which the *main objective or effect is to change, or in some way influence, the reader's opinion or conclusion.* Maps are also used as instruments of propaganda, a subgroup of persuasive maps. While there are many examples from

World War II and the Cold War, propaganda maps have existed throughout history and continue to be made today.

User Limitations

We must not ignore the map user in this discussion of limitations. A user who lacks map reading skills may misinterpret a map or "read into" the map information that can't be learned from the map alone by projecting his or her stereotypes and misinformation onto the map. A common mistake is using a map for the wrong purpose. Most maps are designed for specific purposes or jobs. For example, one cannot get much climate information from a general-purpose map or road information from a weather map.

Significance of Map Limitations to Map Reading, Analysis, and Interpretation

Why is understanding map limitations important to map users? The more knowledgeable a user is about maps, the better his or her interpretation of them. Maps are powerful tools, but they are not intuitive. Misreading a map, even a GPS map, can result in getting lost, or worse, when navigating a boat, plane, car, bicycle, or trail; agencies can allocate funds inappropriately, poor choices of proposed routes can be made, and one can gain erroneous impressions of places—their sizes, their people, their nature. The limitations will be part of the following discussions of different map types and their uses.

MAP READING, MAP ANALYSIS, MAP INTERPRETATION

The terms *map reading, map analysis,* and *map interpretation* are often used interchangeably, but they are not synonymous. I define and use the terms in this book in the following way: Map reading refers to the most basic aspects of map use: finding locations, recognizing symbols and what they stand for, and rudimentary way-finding. Map analysis involves calculations with maps, such as determining steepness of slope and gradient, computing areas, and drawing profiles. Map interpretation is a high-level map use that may involve more than one map to determine the nature of an area or distribution. It requires recognizing and describing spatial patterns, relating and correlating geographic patterns, and essentially bringing all of the available map information together to study a place or subject.

FURTHER READING

Dorling, Daniel, and David Fairbairn. (1997). *Mapping: Ways of Representing the World.* Essex, UK: Prentice Hall.

Garfield, Simon. (2012). *On the Map: A Mind-Expanding Exploration of the Way the World Looks.* New York: Gotham Books.

Greenhood, David. (1964). *Mapping.* Chicago: University of Chicago Press.

Jennings, Ken. (2011). *Maphead*. New York: Scribner.

MacEachren, Alan. (1994). *Some Truth with Maps: A Primer on Symbolization and Design*. Washington, DC: Association of American Geographers.

MacEachren, Alan. (2004). *How Maps Work: Representation, Visualization, and Design* (2nd ed.). New York: Guilford Press.

Monmonier, Mark. (1995). *Drawing the Line*. New York: Henry Holt.

Monmonier, Mark. (1996). *How to Lie with Maps* (2nd ed.). Chicago: University of Chicago Press.

Monmonier, Mark, and George Schnell. (1988). *Map Appreciation*. Englewood Cliffs, NJ: Prentice-Hall.

Parker, Mike. (2009). *Map Addict*. London: Collins.

Tyner, Judith. (2010). *Principles of Map Design*. New York: Guilford Press.

Wood, Denis. (1992). *The Power of Maps*. New York: Guilford Press.

Wood, Denis. (2010). *Rethinking the Power of Maps*. New York: Guilford Press.

Wright, John K. (1942). "Mapmakers Are Human: Comments on the Subjective in Maps." *The Geographical Review*, 23(4), 527–544. [Reprinted in Martin Dodge, Rob Kitchin, and Chris Perkins (Eds.). (2011). *The Map Reader: Theories of Mapping Practice and Cartographic Representation*. West Sussex, UK: Wiley-Blackwell.]

CHAPTER 2

Landmarks of Mapmaking

As the invention of tools is epochal in human history, the invention of the map, which is probably the first intellectual tool, is pre-eminent in human development.

—*Mapping* (p. xiii)

It is difficult to condense the history of mapmaking (cartography) into one chapter. The history of cartography represents, in a sense, the history of the world.[1] In 1977 David Woodward and Brian Harley began the monumental task of creating a multivolume history of cartography. When completed, it will be six volumes. However, Volume 2 is divided into three books, thus making a total of eight books covering the history of cartography through the 20th century. Obviously, in this brief space we can only hit some highlights.

In addition to the enormity of the task, there is a basic problem that faces the researcher studying the history of any type of artifact—namely, gaps in the record. The farther back in history, the fewer maps exist to study. The reasons for this sparsity are many. Maps are frequently created on fragile materials, such as paper, papyrus, silk, or clay. These materials deteriorate or break. When they are created on more durable materials such as copper, silver, or gold, the value of those metals makes the maps vulnerable. There is a story of a 16th-century map of the moon engraved on copper being melted down to make a tea kettle. Maps often have commercial or strategic value, such as to show a shorter route to the Indies or to describe military

---

[1] We must distinguish between history of cartography or cartographic history and historical cartography. History of cartography is the history of maps and mapmaking; historical cartography is the modern creation of maps of historical subjects, such as the Civil War, Ancient Greece, and the Bible Lands.

plans. If the map is in danger of falling into competitor or enemy hands, it may be destroyed. A problem for many more recent maps is lack of currency. There is a vulnerable period in the life of a map when it is too dated to be useful but too new to be considered collectible, and thus it may be discarded. Road maps of the 1930s–1980s were given away at gasoline stations; every year new free maps were available, and the motorist simply tossed the old one in the trash. As a result, surviving maps from the early and middle 20th century are now valued by collectors. Some maps, especially those of preliterate peoples may not be recognized as maps at all; some rock carvings that are now considered maps and the sophisticated stick charts used for navigation by the Marshall Islanders didn't fit early definitions of maps. Archaeological and anthropological studies may unearth more early map forms not yet known.

APPROACHES TO THE STUDY OF OLD MAPS

Just as there are many ways to approach the study of world history, there are many ways to study the history of cartography (see Table 2.1). These approaches or *paradigms* are a framework for studying old maps. Until the last half of the 20th century, history of cartography was largely the domain of map collectors and librarians. One approach they took, which is a common early approach, was simply a *chronological* one that looked at maps from a particular time period. A second approach, favored by collectors, treats maps as decorative objects that Monmonier and Schnell called the "old is beautiful" paradigm. For the most part, this view assumes that maps lacking decorative elements are sterile and unattractive. In general, maps after 1800 were not considered worthy of study. However, with the exception of some collectors, this approach has fallen out of favor. Another aspect of early studies, followed by histo-

TABLE 2.1. Approaches to Studying the History of Cartography

Chronological
Studies maps from a specific time period or through time.

"Old is beautiful"
Looks at maps as decorative objects; not interested in maps after about 1800.

Maps as documents
Interested in the information content of maps; what they show, not how they show information.

Darwinian
Evolutionary approach; assumes that newer maps are better than old maps.

Nationalist
Studies the maps made in a single country.

Regional
Studies the maps of a particular area regardless of where made.

Biographical
Studies the work of a particular cartographer.

Thematic
Studies a particular type of map, such as topographic maps, weather maps, and newsmaps.

Technology
Studies maps made using a particular technology, such as wax engraving, lithography, computers.

Deconstruction
Looks at maps as discourse, based on writings of Jacques Derrida.

Visual rhetoric
Studies how visual images communicate. In the case of maps, looks at their meaning.

Material culture
Studies maps as objects and their significance to culture.

rians, assessed the "information content" of maps, that is, what the map showed. This approach treats the map as an historical document that provides information about places in the past and allows the historian to assess the geographic knowledge of a period. The researcher is not concerned with the map itself, its form, or how it represents places, but only its information and perceived accuracy.

The *Darwinian* paradigm assumes that modern maps are more accurate than older maps; that maps have evolved from simple maps by preliterate peoples to sophisticated computer-generated maps of the 21st century. Although this is the tacit approach in many histories of cartography, it has serious flaws. Sophisticated navigation maps of Marshall Islanders, as well as some crude computer-generated maps, made it clear that this approach is not uniformly viable.

A common method of studying maps is the *nationalist* approach in which the maps made by one nation are examined regardless of what they show. Thus, one might study Dutch cartography of the 15th and 16th centuries, or Arabic cartography of the Middle Ages. One can also take a *regional* approach and look at maps of a particular place or region, such as maps of North America or maps of Washington, DC, regardless of who made them, when, or where.

A number of studies have been made using the *biographical* approach, in which the maps and life of one cartographer are studied. Numerous books have been written about Mercator, for example. One can also take a *thematic* approach and study maps of one type, such as topographic maps, star charts, or medical maps, or a *technology* approach that looks at the way in which some maps were made and the tools of mapmaking. David Woodward used the technology approach in *The All American Map,* which looked at the use of wax engraving on American maps of the late 19th and early 20th centuries. J. H. Andrews examined various technological aspects of map creation, including surveying, in *Maps in Those Days.*

More recently, scholars have examined maps as objects of *material culture* and as *visual rhetoric*, which examine how and what maps communicate, and *deconstruction* of maps, which treats maps as texts to be analyzed.

In studying a map (or maps), it must be put into context; it cannot be viewed in isolation. Maps are made at a specific time and place and within a particular culture. Figure 2.1 shows the "tapestry" of the history of cartography and the way that various disciplines interact in that study.

LANDMARKS IN MAPPING

Arthur Robinson, in *Early Thematic Mapping in the History of Cartography*, used the phrase "revolutions in cartography" to distinguish periods of major change from periods of little change. While his approach assumed the (now much criticized) Darwinian approach, we can identify periods when major changes (for better or worse) take place; this refinement doesn't assume that the maps are better, just that these are periods of change. These "revolutions" come about when three factors are present: a technological change, an increase in data, and a shift in the intellectual aspects of mapmaking, that is, changing paradigms. For example, we can look at the period beginning in the 1960s as such a revolutionary period brought about by satellites, computers, and new map types based on those technologies. The periods between revolutions are primarily times of refinement of the new techniques and beliefs.

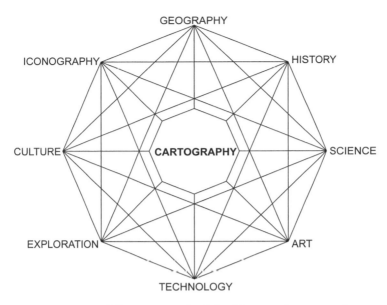

FIGURE 2.1. The "tapestry" of the history of cartography.

Although there are flaws in this method, it is a useful approach for beginning a study of the history of maps and mapping.

For this section, I will look primarily at landmarks in cartography in order to present an overall view of the field. Because I can only touch on some tantalizing high points, the reader might want to look at some of the general histories of cartography listed in the Selected Readings or some more specific sources in the Bibliography. In addition, I will look briefly at the history of some map types and symbols in the relevant chapters.

We have no idea when the first map was made. It appears that the mapping impulse is deeply ingrained in humans, and it is likely that the first maps were scratched in dirt to show directions or were drawn in charcoal from a campfire. There is a strong belief that maps predate writing. There are almost no societies without maps of some form.

The oldest maps still in existence are clay tablets with cuneiform writing from Mesopotamia that show a city, a region, and the known world. The regional map is thought to date from 2500 B.C.E.; the world map (Plate 2.1) is thought to date from about 700–500 B.C.E. An intriguing map carved in stone is the Bedolina map from northern Italy. It has been called the oldest plan of a settlement and is believed to date to the Iron Age, the eighth century B.C.E. (Figure 2.2).

The earliest periods in mapping in both the Western world and China involved understanding the earth's shape and size. Contrary to popular belief, the essentially spherical shape of the earth had been recognized by the Greeks as early as the fourth century B.C.E. By the second century B.C.E., Crates of Mellos had created a globe. The globe was more of a model that posited four basically symmetrical continents, two in the northern hemisphere and two in the south. Only one of those continents, Europe, was based on actual knowledge (Figure 2.3). Eratosthenes (276–196 B.C.E.), who was the head of the Library at Alexandria, discovered a simple way to measure

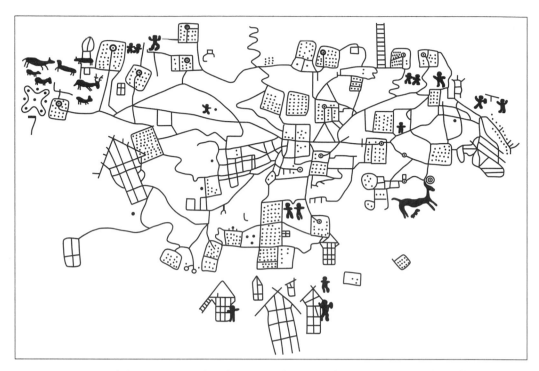

FIGURE 2.2. Bedolina pictograph. The original is carved in stone. From The Oldest Known Plan of an Inhabited Site Dating from the Bronze Age, about the Middle of the 2nd Millennium B.C., *Imago Mundi, 18* (1964) by Walter Blumer. Reprinted by permission of Taylor & Francis.

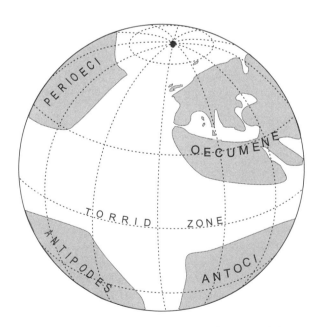

FIGURE 2.3. Reconstruction of Crates globe.

the earth using shadows and determined its circumference to be about 25,000 miles (Figure 2.4). Unfortunately, when Columbus made his first voyage in 1492, he didn't use Eratosthenes' figure, but rather the estimate of Posidonius which had been used in later writings. That figure showed the circumference to be only three-quarters of the correct size. This is suggested as one of the reasons Columbus believed he had arrived in the Indies when he reached the New World.

Claudius Ptolemy (fl second century C.E.), who is generally considered the greatest of the early geographers, like Eratosthenes, was librarian at Alexandria. His major contribution to cartography was a book known as the *Geographia*, which included instructions for making map projections, for making world and sectional maps, and for using latitude and longitude, and which also listed coordinates for about 8,000 places (Figure 2.5). It was here that Posidonius' measurement was written. A number of translations of Ptolemy's work are available.

The Middle Ages, usually placed between the 5th and 15th centuries C.E., are marked by some declines in European cartography and advances in the Arab world and China. European maps are categorized as being of three different types: *Mappae Mundi*, or maps of the world; itineraries; and navigation charts. *Mappae Mundi* were conceptual representations of the world based on biblical beliefs; in their most basic form they are circular and show three continents, Europe, Asia, and Africa, surrounded by an ocean. They are frequently referred to as T-in-O maps because of their shape. On T-in-O maps, East (the Orient) is at the top of the map and Jerusalem is at the center; a circumfluent ocean forms the O, and the T is formed by the Don and Nile rivers and the Mediterranean Sea; the earth is presumed to be flat (Figure 2.6).

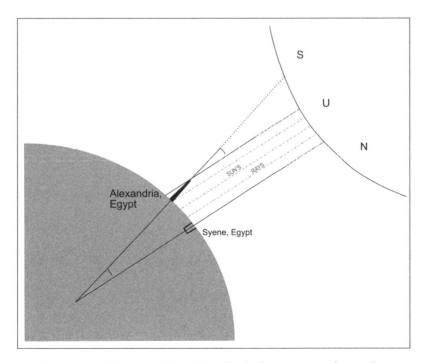

FIGURE 2.4. Eratosthenes' method of measuring the earth.

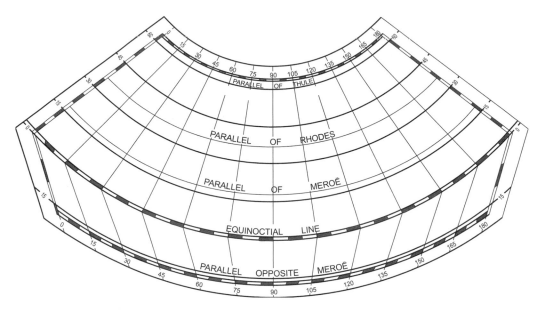

FIGURE 2.5. Ptolemaic world projection. From *Map Making* by Lloyd A. Brown. Copyright © 1958 by Lloyd A. Brown. Adapted by permission of Little, Brown and Company. All rights reserved.

The size of *Mappae Mundi* range from small enough to fit in the palm of one's hand to about 12 feet (3.6 m). The best known of the large maps still in existence is the Hereford map located in the Hereford Cathedral in England (Plate 2.2). It was drawn in about 1300 on a single sheet of calfskin vellum and measures 64 inches by 52 inches (1.58 × 1.33 m). The Ebsdorf map, which was made in Germany at about

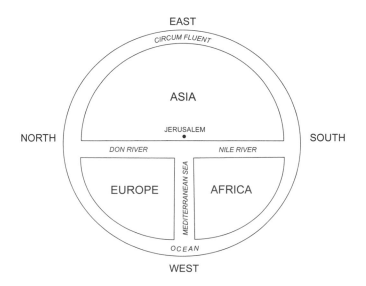

FIGURE 2.6. Schematic diagram of a T-in-O *Mappa Mundi.*

the same time, was made on 30 pieces of goatskin and was about 12 feet by 12 feet (3.65 m). Tragically, the Ebsdorf map was destroyed in 1943 during an air raid. Fortunately, however, a facsimile had been made in the late 19th century. The Hereford map can be viewed at the cathedral and at the Hereford Cathedral website listed in Resources at the end of this chapter. *Mappae Mundi* are concepts of the world and contain much myth and conjecture; in terms of cartographic and geographic knowledge, these are usually considered a step backward.

The second type of medieval map is *the itinerary* or route map. The most famous of these maps is the Peutinger Table which in its original form was over 20 feet long and 1 foot wide. It is believed to be based on a Roman itinerary of the first century C.E., but mostly fourth-century and some 16th-century additions. Matthew Paris, a monk in England, produced an itinerary map of England for 13th-century travelers. These itinerary maps were strip maps that showed pilgrimage routes in a long, diagrammatic form, much like a modern triptik.

The third type of medieval map was the *Portolan chart* (Plate 2.3). These were sailing charts that developed sometime after the magnetic compass; this compass was first used by the Chinese and was introduced into Europe in the 13th century. The oldest Portolan chart is the *Carte Pisane*, created in about 1290. A typical Portolan shows coastlines around the Mediterranean Sea crisscrossed by straight lines of constant compass direction (rhumb lines) with compass roses at intersections of the line. These charts made it possible to sail a course with only a compass as a navigational aid. Columbus would have used such a chart when sailing the Mediterranean Sea.

Arab cartography had its origins in roughly the ninth century C.E. The Arabs' early works were based on Ptolemy, and they created celestial maps; a celestial globe still exists. In part, Arab mapping was to aid in determining the direction of Mecca. Some of their world maps resemble the T-in-O maps of the Europeans. Arab cartography has been described as having three phases: an early phase based on the work of Ptolemy, a middle phase that is distinctly Islamic, and a third phase that is based on the maps of Al-Idrisi who was in the court of the Norman King Roger of Sicily in the 12th century C.E. These phases are not strictly chronological, and at times all three coexisted.

As with maps of the early Western world, our knowledge of Asian maps is sketchy, and the majority of that information is based on a few examples, primarily Chinese, and writings about them. Surveys of China were made in the 6th century B.C.E., but the maps that probably accompanied these works do not remain.

The Chinese are credited with the earliest printed map in about 1155 C.E., which predates printed maps in Europe by 300 years, and coincides with the use of the magnetic compass in the 11th century and the invention of paper in the second century C.E. About the same time that Ptolemy was writing, Chang Heng, an astronomer, introduced a rectangular grid.

In 1973, maps drawn on silk were discovered in a tomb from the second century B.C.E. in Hunan Province. One of these maps was a topographic map that showed a part of the province with streams, mountains, and roads; another was a military map that represented the relative military importance of various features; the third map was of a prefecture seat. These maps are quite sophisticated in their symbolization. All had south at the top.

By the 15th century, major changes were taking place in technology, information, and geographic thinking. Printing with moveable type was invented, which made possible the creation of multiple exact copies of a map and the more rapid dissemination of books and maps; the first printed European world map was a small T-in-O map in 1472. The works of Ptolemy, which had been essentially forgotten in Europe but were known in the Arab world, were translated into Latin in the early part of the century, and Ptolemaic maps rapidly replaced T-in-O maps. The first versions were hand-lettered manuscript versions, but they were quickly followed by printed copies.

In addition to the impact of printing, voyages of exploration were carried out by Europeans and Chinese. Prince Henry of Portugal initiated a large number of these explorations, and, of course, Columbus made the first of his voyages to the New World under the auspices of Spain. The oldest surviving European globe, by Martin Behaim of Nuremburg, was created in 1492 but does not show Columbus's discoveries since it was made at the time of the voyage. During the 16th century, many projections for world maps were devised, including Mercator's cylindrical projection (see Chapter 5). Voyages of exploration continued throughout the 16th and 17th centuries. Although they were largely designed for conquest and enrichment of the sponsoring countries, they did bring back much new geographical information. This, in part, led to the creation of large atlases. During the late 16th and early 17th centuries, the Low Countries—Netherlands and Belgium—produced landmark atlases. Mercator, Hondius, and Ortelius were major names in the "golden age of atlases."

By the end of the 17th century, which marked the beginnings of the Scientific Revolution and the Enlightenment, the Royal Society had been formed in Britain, and similar societies were organized in other European countries. One aim of the Royal Society was expressed in "Directions for Seamen Bound for Far Voyages," which encouraged sailors and captains to observe people, flora and fauna, and to keep careful notes of the lands they visited. One who did so was William Dampier, whose observations were published as *A New Voyage Round the World* in 1697. Not only was there interest in the earth, but astronomers turned the newly discovered (1608) telescope on the moon and by the middle of the 17th century, sophisticated maps of the moon were created.

The 17th and 18th centuries were marked by increased cadastral (property), route, hydrographic, and topographic mapping and the beginnings of thematic mapping. Surveys to measure the size of the earth were instituted, such as that by Jacques Cassini in 1713, which resulted in more accurate maps. Other surveys led to large-scale topographic mapping. The invention of the chronometer, which made possible the determination of longitude at sea, was a major development for navigation (see Chapter 4). Collecting demographic data by regular censuses began in the middle of the 18th century, with Sweden in 1749, the United States in 1790, and Britain in 1801; these censuses, plus the development of the discipline of statistics, led to thematic maps of population. Although a few thematic maps showed such things as winds and lines of magnetic declination earlier, thematic cartography became a major part of the field of cartography in the 19th century. Remote sensing, in the form of aerial photography, traces its origins to the middle of the 19th century.

As noted earlier, the middle of the 20th century marks a major revolution in cartography. The development of computers and satellites led to more sophisticated remote sensing that has provided a wealth of data. These also resulted in the development of GPS and GIS for creating maps. Scholars in cartography now began to look at maps in different ways; rather than being seen merely as repositories of information, maps were viewed as instruments of communication and visualization. Researchers analyzed how maps work and studied the power of maps.

In the following chapters we will look not only at how to read and interpret maps, but also at the development of various map types.

FURTHER READING

Andrews, J. H. (2009). *Maps in Those Days: Cartographic Methods before 1850*. Dublin: Four Courts Press.

Bagrow, Leo, and Raleigh Skelton. (1966). *The History of Cartography*. Cambridge, MA: Harvard University Press.

Brown, Lloyd. (1949). *The Story of Maps*. New York: Bonanza Books.

Hoogvliet, Margriet. (1996). The Mystery of the Makers: Did Nuns Make the Ebsdorf Map? *Mercator's World*, 1(6), 6–21.

Monmonier, Mark, and George Schnell. (1988). *Map Appreciation*. Englewood Cliffs, NJ: Prentice-Hall.

Robinson, Arthur H. (1982). *Early Thematic Mapping in the History of Cartography*. Chicago: University of Chicago Press.

Thrower, Norman J. W. (2008). *Maps and Civilization* (3rd ed.). Chicago: University of Chicago.

Wilford, John Noble. (2000). *The Mapmakers* (revised ed.). New York: Vintage Books.

History of Cartography Project

The History of Cartography, Volumes 1–6. (Volumes 1, 2, and 3 are viewable online.)

Volume 1. (1987). *Cartography in Prehistoric, Ancient, and Medieval Europe and the Mediterranean.*

Volume 2, Book 1. (1992). *Cartography in the Traditional Islamic and South Asian Societies.*

Volume 2, Book 2. (1995). *Cartography in the Traditional East and Southeast Asian Societies.*

Volume 2, Book 3. (1998). *Cartography in the Traditional African, American, Arctic, Australian, and Pacific Societies.*

Volume 3. (2007). *Cartography in the European Renaissance.*

Volume 4. *Cartography in the European Enlightenment* (forthcoming).

Volume 5. *Cartography in the Nineteenth Century* (forthcoming).

Volume 6. *Cartography in the Twentieth Century* (forthcoming).

RESOURCES

James Ford Bell Library
   www.lib.umn.edu/bell/map
British Library Map Collection
   www.bl.uk/onlinegallery/onlineex/mapsviews

Hereford Cathedral
   www.herefordcathedral.org/visit-us/mappa-mundi-1
History of Cartography Project
   www.press.uchicago.edu/books/HOC/index.html
   Volumes 1, 2, and 3 of *The History of Cartography*
Library of Congress Map Collection
   memory.loc.gov/ammem/gmdhtml
Map History
   www.maphistory.info
   A "gateway" to information and discussions of history of cartography.
Newberry Library, Map Collection
   www.newberry.org/maps-travel-and-exploration
Peutinger Map
   http://peutinger.atlantides.org/map
David Rumsey Map Collection
   www.David Rumsey.com

CHAPTER 3

Map Basics

There is also as much to understand about what a
given map won't do as about what it does.

—*Mapping* (p. x)

While we will be looking at many different map types, there are some tools or skills
that apply to almost all maps. Understanding these is essential to reading, analyzing,
and interpreting maps.

THE PARTS OF A MAP

Maps are made for many different purposes, but they have certain elements in common. Figure 3.1 shows a simple map layout, but it must be kept in mind that not all
maps contain each of these elements and there is no standard location for the elements. They may vary according to the cartographer's purpose, sense of design, and
the publisher's or client's requirements.

- *Subject area.* This is *the* map, the subject and purpose of the page.
- *Title.* Just like books and articles, maps have titles. Ideally, the title includes the
location of the map and the subject, such as the population of New York or rainfall in
Australia and may also include a date, such as the population of New York in 2010.
- *Legend.* The legend, which is sometimes called the *key*, explains the symbols
used on the map. Most of the time this is included on the map page, but for some
types of map, a separate key sheet is available. This is most often the case with map
series, such as topographic maps or geologic maps.

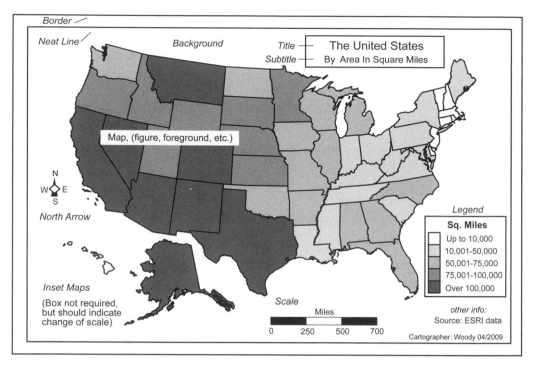

FIGURE 3.1. Elements of a map.

• *Scale.* The scale of the map compares the size of features on the map to their actual size. It may be expressed numerically as a ratio (1:24,000), in words (1 inch represents 2,000 feet, 1 centimeter represents 240 meters), or graphically (as a line marked in units). Scale will be discussed in detail later in this chapter.

• *Orientation.* Contrary to popular belief, maps do not always have north at the top, but it is a common convention, so if there is no other indication it is usually assumed that north is at the top. Maps in the Middle Ages placed east at the top, and some early Chinese maps put south at the top. Orientation is shown in three ways: latitude and longitude, discussed in the next section, a north arrow that points toward north or a compass rose that shows the cardinal directions, N, S, E, W.

• A map may have an *inset map*, which is a small map included within the frame of the main map. The inset may be used to focus in on an area, to allow a larger map on the page by moving some elements, or to put the area in a larger context (Figure 3.2a, b, c). This is frequently done with maps of the United States on which Alaska and Hawaii are included as insets rather than in their proper location. This arrangement has caused some problems, especially when used on school maps, because many children assume Alaska is an island. The insets are also not usually the same scale as the main map; if the purpose is to focus on an area, the inset will "zoom in" on the area.

• There may be other *explanatory materials* on the map. A world map may show which projection was used to construct the map; a map showing statistical information may show the source of the data; just as in books and articles, the map author, publisher, and the date the map was made may be shown.

• The page itself has several elements. A line may separate the map's subject area from the rest of the page. This line is called the *neat line*; there may also be a separate *border* beyond the neat line; and finally, beyond the border is the *margin*. Sometimes explanatory materials are included within the margin.

THERE'S A MAP FOR THAT

As Greenhood says, "There is no walk of life without a guide in the form of some kind of map" (*Mapping*, p. xiii). In this section we provide a brief overview of the many kinds of maps that are available, and they will be discussed more thoroughly in individual chapters (see Table 3.1).

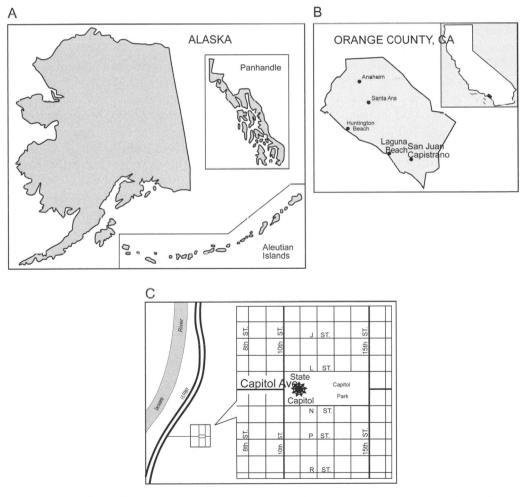

FIGURE 3.2a, b, c. Inset maps may be used to get a larger map on the page (gain scale), to focus on a small area, or to show where a place is in a larger context.

TABLE 3.1. Map Types

General/reference maps Show general nature of an area, but do not emphasize any one feature.	*Special-purpose maps* Designed for specific uses and users, such as geology, weather maps, and navigation maps.
Topographic maps General, but emphasize terrain, small and medium scale.	*Historical maps* Modern maps that show historic events.
Cadastral maps Show property lines.	*Navigation maps* Aeronautical charts, nautical charts, road maps, and bicycle maps.
Thematic maps Show a particular topic, such as climate, population, and agriculture.	*Astronomical maps* Maps of the heavens, maps of planets.

First, we must realize that there is no one general, all-purpose map. Maps are created for specific reasons, to show specific topics or features. Not using the right map for the job will, at best, leave the reader with insufficient information and, at worst, will mislead. Here is a simplistic example: A map of geologic features will not show many, if any, cultural features.

The closest to an all-purpose map is the *general map* or reference map. Such maps are found as individual maps, and in atlases as maps of regions, states, or countries. No one feature is emphasized; they show transportation, rivers, towns, terrain, and boundaries. Not all features are shown, however. For example, some general maps show roads and railroads, others only roads. Because these maps are often small scale (showing a large area but little detail), they are quite *generalized* and thus, only major roads or towns over a specific size might be shown. For more detailed information, the reader will need to consult other larger-scale maps.

A variation of the general map is the *topographic map* (see Plate 12.7 in Chapter 12). While we usually think of topographic maps as maps that show terrain, the *topography* of an area actually refers to the natural and human-made features of a place. In the United States, topography often is considered to include only the natural features. Topographic maps, however, show both cultural and natural features. Elevations are shown by means of *contour lines* (see below) on modern maps. Vegetation categories, water features, cultural features, such as buildings, roads, railroads, trails, and canals are all shown. Most major countries produce topographic maps of their country, and normally they are produced by government agencies. Thus, in the United States, the U.S. Geological Society has produced topographic maps since 1879; in England the Ordnance Survey is responsible; Switzerland's maps are produced by the Federal Office of Topography (SwissTopo); and Germany's are produced by the Bundesamt für Kartographie und Geodäsie (BKG). Topographic maps are usually considered large or medium scale, and mapping agencies may produce two or more *series* of maps based on their scale. Appendix D contains a list of topographic mapping producers. Chapter 8 treats topographic maps in detail.

Cadastral maps are one of the oldest types of map; they are maps of property and are used for documenting land ownership and for tax purposes. One can learn much from cadastral maps depending on when or where they were made. They are of special interest to genealogists and historians because the names of property owners are often on the map. They can indicate when and by whom an area was settled because different countries have different cadastral systems. Cadastral maps are covered in Chapter 4.

Thematic maps illustrate a particular subject or theme, such as population density, election results, or vegetation. At one time these were called *statistical maps*, but since many are qualitative in nature, the term *thematic* is now preferred. Often these are small-scale maps and show little information other than the theme. Some base information, such as state or county boundaries and major cities, may be shown for reference (Figure 3.3).

Many map types for specific subjects are grouped under *special-purpose maps*. Here are found large-scale geologic maps for studying the geology of an area; weather maps showing barometric pressure, fronts, wind direction, and the like. Weather maps are not to be confused with thematic maps that show climates for an area. Weather maps are designed for specialists in a field, although simplified weather maps are a common feature of the newspaper or TV news show. Historical maps, recreation area maps, and maps of tourist attractions all come under this heading; they are designed for specific users and purposes.

One widely used map type is the *historical map*. As we noted in Chapter 2, there is a difference between historic maps and historical maps. Historic maps are old maps made at a particular time; historical maps are modern maps made to represent a historical time. Thus, a map made during the Civil War is an historic map, but a map created in the 21st century to show Civil War battlefields is an historical map.

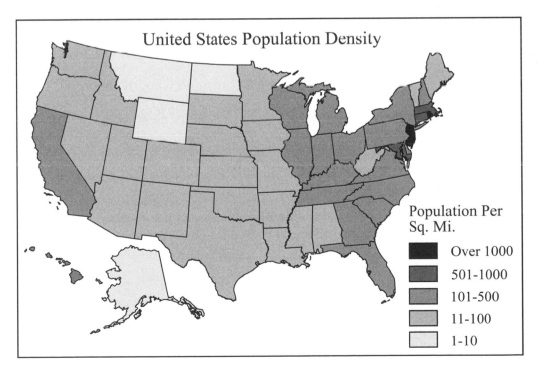

FIGURE 3.3. Thematic maps illustrate a particular subject.

Historians make use of both kinds; they use historic maps as documents in research and illustrate their research with historical maps.

The next groups of map types could also be considered special purpose, but they are so widely used that we will treat them separately.

Navigation maps include *nautical charts, aeronautical charts, road maps,* and even *bicycle maps.* All are now used or can be used in conjunction with GPS. Navigation maps are used for way-finding. Nautical and aeronautical charts are large-scale maps used for water and air navigation, respectively; to be used effectively they require special map reading skills. Road maps are probably the most familiar map type for most people. Whether a paper map, an online map, or a map on an automobile GPS unit, they are widely used on an everyday basis.

Star charts and *planetary maps.* Mapping and map reading are not confined to earth. For centuries astronomers have created maps of the heavens, and since the 17th century they have created maps of the moon and Mars. Since the advent of man-made satellites and manned and unmanned space flights, more sophisticated maps of the moon and Mars have been created. We now have geologic maps of both bodies.

Atlases are not a map type, but a collection of maps of the world, a region, or even a theme. A typical world atlas contains maps of various countries or regions, usually at a uniform scale, and many thematic maps for the world and regions. One can learn a great deal about a place just from an atlas—location, population, climate, economy, vegetation, and the like can all be determined. Some atlases deal with only one area, such as California.

There are thematic atlases that illustrate the history of an area or of specific events, or politics, or women's issues. Thus, there are atlases of the Civil War and of World War II. A popular type of atlas is a *road atlas* that contains maps of individual states, cities, and the entire country, focused on the road networks. These are useful in planning multistate trips. There are also topographic atlases that show contour lines for elevations.

SYMBOLS AND SYMBOLIZATION

Symbols are just the mapmaker's shorthand.

—*Mapping* (p. 185)

All maps are symbolized; that is, marks on the map represent features, and symbols have sometimes been described as the language of maps. Each map has a set of symbols, which might be illustrated in the legend of the map. For some kinds of maps, especially topographic maps, a separate symbol sheet might be available.

Only three kinds of marks can be put on flat paper—points, lines, and areas—and so symbols are described as point, line, or area symbols. Symbols can also name or illustrate some quality of what they represent or show the quantity represented; thus, symbols are qualitative or quantitative. Quantitative symbols may allow the reader to determine exact values, or they may be placed in categories such as small, medium, or large. The cartographer uses a variety of different elements called *graphic variables* or *visual variables* to create symbols. These variables are shape, size, tonal value (shades), hue, pattern/texture, and orientation (Figure 3.4). For the visual vari-

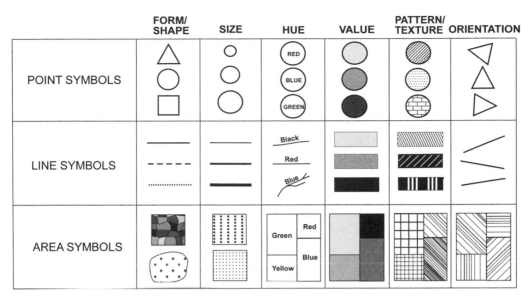

FIGURE 3.4. The visual variables.

able of shape or form, symbols might be pictorial, purely abstract, or geometric, or something in between, called associative, that gives the "feel" of what is represented (Figure 3.5). Thus, an almost infinite number of symbols can be created.

When we look at specific map types, we will treat the symbols used on those maps in more detail, but here I will present a brief overview of some of the most common symbols.

	PICTORIAL	ASSOCIATIVE	ABSTRACT
POINT SYMBOLS	School / Mine	Mountain Peak / Building	City / Airport
LINE SYMBOLS	Highway / Railroad / River	Disputed Boundary	Railroad / Trail
AREA SYMBOLS	Coniferous Forest / Grassland	Woodland / Limestone	Forest / Crops

FIGURE 3.5. Shapes can be pictorial, associative, or abstract.

Qualitative Symbols

Point Symbols

Point symbols occur at a point and in the simplest form give data only for that point. Common examples are a dot showing the location of a town, or a square showing a building. They show only location and the type of feature, not the population of the town or the height of the building. These are *qualitative point symbols* (Figure 3.6).

Linear Symbols

The lines shown in Figure 3.7 show rivers, roads, and boundaries. The line thickness doesn't vary, and there is no indication of size. Only the location, the nature of the feature, and the pattern of the feature can be determined. These are *qualitative line symbols*.

Areal Symbols

Like other qualitative symbols, qualitative areal symbols show what is found in an area such as a field, county, country, or state (Figure 3.8). Thus, a field may show corn or soybeans, or a region might be grassland or forest. Because the symbol covers the actual area, it is also possible to determine the size of the field or the extent of the forest.

Quantitative Symbols

Point Symbols

Simple dot maps use uniform symbols (they may be circular, rectangular, triangular, or other shapes), each of which represents a given quantity, such as one dot for 500 people. The cartographer places the dots in the centers of distribution within the area, thus revealing the pattern of distribution (Figure 3.9). *Dot density maps* are similar in that one dot represents a specific quantity, but instead of being placed in

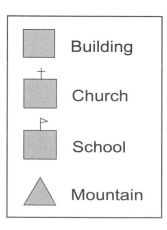

FIGURE 3.6. Qualitative point symbols.

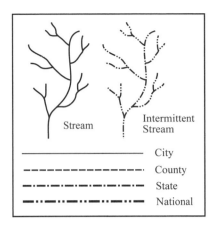

FIGURE 3.7. Qualitative linear symbols.

the centers of distribution, they are placed randomly within the areas. Patterns of distribution cannot be determined unless the enumeration areas are quite small.

Graduated circle maps employ circles of varying sizes to represent quantities (Figure 3.10). If the circle represents a point location, such as a city, it is placed at that point; if it represents an area, it is usually placed in the center of the area—for example, the population of a state. Graduated circles may show either actual amounts or categories. The circles are drawn so that the *area* of the circle, not the radius or diameter, is proportional to the amount represented. When categories are represented, a set number of circles are shown in the legend with amounts such as 1–1,000, 1,001–5,000, 5,001–10,000, and the like. For a truly proportional circle,

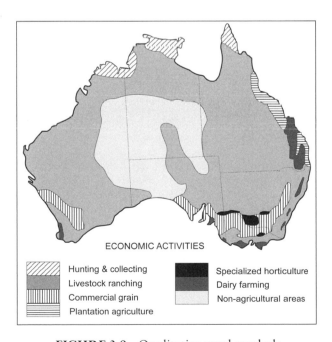

FIGURE 3.8. Qualitative areal symbols.

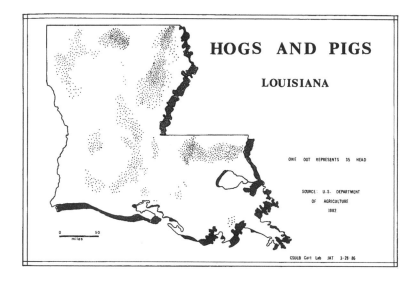

FIGURE 3.9. Dot map.

representative circles are shown with a scale device. Dot and graduated circle maps are discussed in detail in Chapter 9.

Linear Symbols

For quantitative line symbols, the width of the lines vary, showing the size of the river, the classification of the road, the volume of traffic, or the significance of the bound-

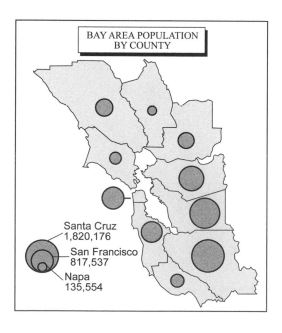

FIGURE 3.10. Graduated circle map.

ary. An early quantitative line symbol is the *flow line* that was first used in the 19th century. Flow lines usually show direction and amount, such as traffic in the south-bound interstate, and the lines vary in width with the amount shown (Figure 3.11).

Isarithmic lines are one of the most widely used types of symbol. Weather maps, topographic maps, and population maps all make use of types of isarithmic lines. An isarithmic line is a line that joins all points having the same value above or below a zero value called the datum (Figure 3.12). *Isoline* and *isogram* are other general terms for these lines, but many different types of isarithms have their own names. A few of the more common ones are:

Isotherm—line of equal temperature

Isobar—line of equal barometric pressure

Isohyet—line of equal rainfall

Isobath line of equal depth below sea level

Isogonic line—line of equal magnetic declination

Isochron—line of equal travel time

Probably the most familiar isarithmic line is the *isohypse*, a line of equal elevation, usually called a *contour line* (Figure 3.13). A contour line joins all points that have the same elevation above or below a datum, usually sea level. They are the most commonly used of the isarithms and are the basis for topographic maps. Contours will be discussed more fully in Terrain Representation later in this chapter and also under thematic maps and topographic maps.

Parallels and meridians are also special forms of isarithms since they join all points having the same latitude and longitude, respectively.

Areal Symbols

Choropleth maps are statistical maps that color or shade areas according to the value or amount of the item being mapped. Because the areas usually vary in size, such as states, to be meaningful choropleth maps must show the statistics as ratios or percentages. For example, in Figure 3.14 each area has 5,000 people, but one area has 10 square miles and the other has 100 square miles. If the areas are shaded simply according to total population, the map will be misleading; if they are shaded accord-

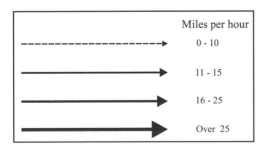

	Miles per hour
- - - - - - - - - - →	0 - 10
———————→	11 - 15
———————→	16 - 25
———————▶	Over 25

FIGURE 3.11. Quantitative linear symbol, flow lines.

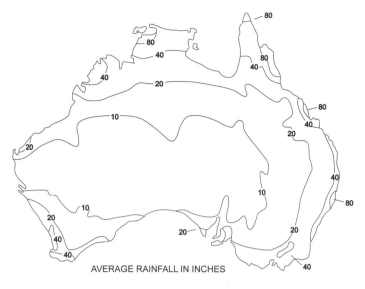

AVERAGE RAINFALL IN INCHES

FIGURE 3.12. Isarithmic map.

ing to population per square mile, the reader can get an idea of the density of population and so the map is much more meaningful and useful.

TERRAIN REPRESENTATION

A special group of symbols and map types are those used for relief or terrain representation. From very early times, showing hills and mountains on flat maps has been

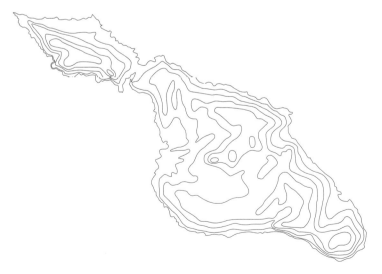

FIGURE 3.13. Contour lines are a type of isarithmic line.

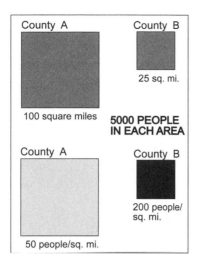

FIGURE 3.14. Choropleth maps are misleading if absolute values are used.

a problem for cartographers, and many methods have been devised. Some symbols are designed to give a general impression of the relief, others to allow measurement of elevation. While some of these symbols are no longer in use, if you are doing a historical study using older maps, it is helpful to understand how they were used and their limitations.

One of the earliest methods is the one you might use on a sketch map, a simple profile drawing of a hill (Figure 3.15). These symbols have been found on the early maps of Mesopotamia and are called the *molehill* or *sugar-loaf* technique. They only give an idea of the terrain and have many disadvantages: Since the symbols are side views, they cannot be in their proper position; that is, they are not *planimetrically correct*, and they block out features "behind them." The appearance of the symbols varies greatly; some mapmakers tried to draw an actual picture of the mountain, whereas others used a stylized symbol. The symbols were not drawn to scale and might even vary within the map. Therefore, the reader had no idea of how big a feature was; hills and mountains often looked the same.

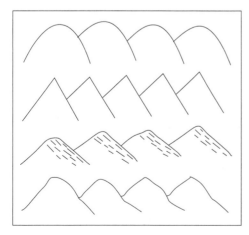

FIGURE 3.15. Molehill or sugar-loaf hill symbols.

   In the late 18th century, a more scientific approach was introduced with *hachures* (Figure 3.16), which are short, straight lines drawn in the direction of the slope. They are planimetrically correct, and spot elevations are shown for measured heights. In the early part of the 19th century, Johann Georg Lehmann, a German cartographer, devised a hachuring system that was based on the slope of the land; the lines were drawn at right angles to the slope, and the thickness of the line corresponded to the steepness of the slope. This highly graphic method was very popular in the 19th century, especially on large-scale maps such as military surveys. During the same period, another hachuring system was devised that had lighter hachures on the north and west and darker hachures on south and east slopes that gave the impression of sunlight and shadow. This variation was used in one of the important atlases of the time, the Dufour atlas of Switzerland.

   While hachures show features in their correct plan position, they have many limitations. Elevations cannot be determined, so a few spot elevations are given. While, on large-scale maps, hachures produce a very attractive map with a good impression of the terrain, at small scales, such as world atlases, they degenerate into what has been variously called "hairy caterpillars" or "woolly worms" (Figure 3.17). In spite of this drawback, they were widely used in atlases until the middle of the 20th century.

   We have already noted that the *contour line*, which is an abstract symbol and a type of isarithm (isohypse), connects points that have the same elevation above a datum (usually mean sea level). Contour lines are planimetrically correct; that is, features are shown in their correct horizontal positions, and elevations between the lines can be estimated or *interpolated*. Although contour lines were actually first used on land maps in the 18th century, because they require exact elevations for many points, they weren't widely used until the late 19th century when major surveys of the land were made. Contour lines are the primary symbol used on topographic maps to show elevation. They will be discussed in more detail in Chapter 9.

FIGURE 3.16. Hachures as used in the 19th century. From *Cartographic Relief Presentation* by Eduard Imhof (1982), Walter de Gruyter GMBH. Reprinted by permission.

FIGURE 3.17. "Woolly worm" hachures.

As with the other symbols here, contour maps do have limitations. Although they are planimetrically correct and elevations can be measured, to an untrained map user the lines may seem confusing. Contour lines do not have a three-dimensional appearance, so an inexperienced reader can't immediately identify hills and valleys. They, like hachures, are best used on large-scale maps; on small-scale maps they must be generalized, and, therefore, on such maps, they are often combined with colors or shading.

Layer tint maps, also called *hypsometric* or *altitude tint* maps, add color or shades of gray to contour lines (Figure 3.18). Different colors are used between the contours to help distinguish elevations. Several different color schemes have been used; each publishing company has its own. In general, the schemes all attempt to give the feeling of a third dimension. The most common is based on the theory that cool colors—blues and greens—appear farther away than warm colors—reds and oranges—and that by using cool colors for low elevations and warm colors for higher elevations there will be an impression of height. Unfortunately, map readers frequently misinterpret the colors. They make three common errors: (1) They assume

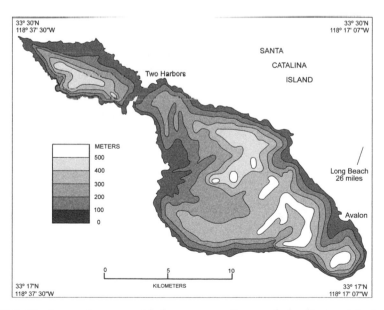

FIGURE 3.18. Layer tints are added to contour lines to help distinguish elevations.

that the colors represent rainfall, vegetation, or temperature. In this case, the reader assumes that low areas shown in green are lush, cool, and damp and brown areas are hot, dry, and desert-like. (2) Because there are sharp lines separating colors, readers may interpret them as a series of plateaus or areas of uniformity. (3) They assume that the lines chosen have some special significance; that they represent boundaries of some sort, whereas they are simply convenient, but arbitrary, intervals.

Relief shading is a way of simulating a third dimension by shadowing and highlighting the area to give the impression of a relief model (Figure 3.19). The map looks as though the land is illuminated from the upper left, normally the northwest. This is the most common method used on atlases today. Relief shading is often used in combination with contour lines.

Digital elevation models (DEMs) are terrain maps created on the computer with satellite data. DEMs give the impression of a three-dimensional surface by the array of elevation values above a datum over an area. They are sometimes presented in shades of gray or may have hypsometric tints added to help in interpreting elevations (Plate 3.1). There is as yet no formal definition of a digital elevation model among scientists in the field.

SCALE

> Give a good map-muser an inch and he knows how to take a
> mile. He never ceases to enjoy the plain but curious fact that
> an inch of paper can be the equivalent of a mile . . . of world.
>
> —*Mapping* (p. 42)

We encounter scale at an early age. A child's dollhouse is a scale model of a house; a toy train is a scale model of a real train; model cars, trucks, and airplanes are all scale models of real-world counterparts. Maps are scale representations of the earth.

Scale is simply the size of a representation compared to the real object expressed as a ratio or fraction. Thus, a model of a truck that has a scale of 1:53 or 1/53 is one fifty-third the size of a real truck. A map with a scale of 1:1,000,000 is one-millionth

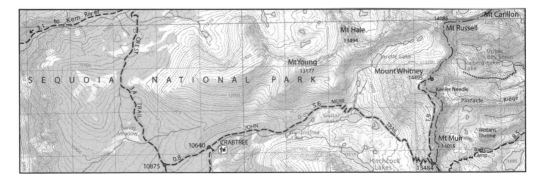

FIGURE 3.19. Relief shading gives the impression of a third dimension to contours. From *www.tomharrisonmaps.com/*. Reprinted courtesy of Tom Harrison.

the size of the earth, and the features shown on the map are one-millionth of their actual sizes. This ratio is called the *representative fraction* (RF), or simply the scale of the map. Notice that no units such as miles or kilometers are attached to the representative fraction. Because of this, the RF can be used regardless of the units of measurement. It means 1 unit on the map represents *x* units in the real world. Thus, in our truck example, 1 inch on the model represents 53 inches on a real truck; for our map example, 1 inch on the map represents 1,000,000 inches on the real world. If we are working with metric units, 1 centimeter on the truck represents 53 centimeters on the real truck, and 1 centimeter on the map represents 1,000,000 centimeters on the earth.

Of course, it is difficult to picture 1,000,000 inches or 1,000,000 centimeters, so we often use a second form of scale, the *verbal scale* (Figure 3.20). Verbal scales are expressed with units so that 1:53 becomes 1 inch represents 53 inches or 4.44 feet and 1 inch on the 1:1,000,000 map represents 15.78 miles or 1 centimeter represents 10 kilometers. When working with inches and feet, the denominator of the RF is divided by the number of inches in a foot; for centimeters and kilometers, the denominator is divided by the number of centimeters in a kilometer. Conversion factors are presented in Appendix C. The verbal scale is often rounded off, so that 1:1,000,000 is *about* 1 inch to 16 miles (in metric it is exactly 10 km). These are usually used for estimates.

On many occasions we need to take a ruler to a map to calculate distances and sizes. We can measure the distance on a 1:1,000,000 map with our ruler, determine that it is 8 inches, and calculate from that a distance of 15.78 × 8, or 126.24 miles. If the map distance between two points is 10 centimeters, we can calculate the distance as 100 kilometers. To eliminate cumbersome calculations, many maps include a *graphic scale* (also called a *linear scale* or *bar scale*). This is a small "ruler" on the map that allows us to measure distances directly (see Figure 3.21).

We often encounter the terms *large scale* and *small scale*. These are relative terms without a specific number attached. Large-scale maps show small areas in great detail, and small-scale maps show large areas in less detail. A city map is a large-scale map, and a world map is a small-scale map. The larger the denominator of the representative fraction, the smaller the scale of the map. This can be thought of as analogous to slices of a pie: 1/4 of a pie is a larger piece than 1/8 of the pie. Within some map series, such as topographic maps, certain scales may be specified as large, medium, and small *for the series* (Figure 3.22).

Because a small-scale map shows a large area on a relatively small page, much

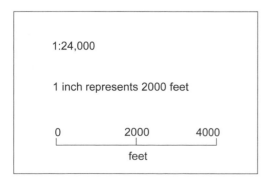

FIGURE 3.20. Methods of showing scale.

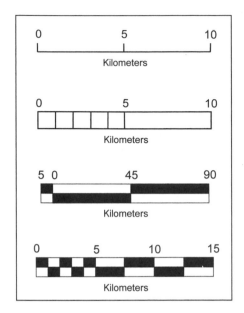

FIGURE 3.21. Graphic or bar scale.

detail must be omitted: for example, coastlines will be smoothed, small towns omitted, or roads shown, but not railroads. This process is called *generalization*; all maps are generalized to some extent, but small-scale maps are highly generalized.

FURTHER READING

Kimmerling, Jon, Aileen Buckley, Phillip Muehrcke, and Juliana Muehrcke. (2010). *Map Use: Reading, Analysis and Interpretation.* Redlands, CA: ESRI Press.

Tyner, Judith. (2010). *Principles of Map Design.* New York: Guilford Press.

Wade, Tasha, and Shelly Sommer (Eds.). (2006). *A to Z GIS.* Redlands, CA: ESRI Press.

RESOURCE

USGS Maps, Imagery, and Publications
www.usgs.gov/pubprod

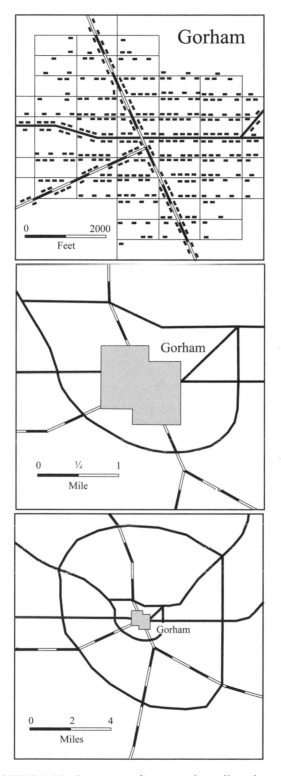

FIGURE 3.22. Large-, medium-, and small-scale maps.

CHAPTER 4

The Figure of the Earth and Coordinate Systems

If the whole world were flat, like our back yard
. . . rectangular coordinates would serve all map
purposes. But with a globular world we have to
develop a set of coordinates which fit it just as
snugly and precisely as a grid fits a flat, rectangular
surface.

—*Mapping* (p. 6)

For this chapter it is helpful to have a globe at hand. It needn't be a large or expensive globe to understand the concepts. Two hundred years ago, globes were considered scientific instruments and were quite expensive, but now they are often relegated to the toy store or used purely for decoration. They are still, however, the most useful tools for understanding the earth's grid and earth–sun relations.

COORDINATE SYSTEMS

Geographic Coordinates

Children learn early that the earth is a ball. More exactly, it is a rather lumpy figure that is close to spherical but is flattened at the *poles* and bulges at the *equator*, and it is called the *geoid* or earth-shaped figure. The diameter of the earth is 7899.98 statute miles (12,713.47 km) at the poles and 7926.68 statute miles (12,713.44 km) at the equator. As represented on a 12-inch (30-cm) globe, the difference is less than the thickness of a sheet of paper, only about 0.0389 inches (0.099 cm). Thus, for most of our purposes, the deviations from a sphere are small enough that it can be considered a sphere.

Children also learn that the earth is spinning and that it goes around the sun. Why is that important? A globe represents the earth; it is a small model of it and spins on an imaginary axis, the ends of which are the *north* and *south poles*. If we draw a line around the spinning sphere midway between the poles, we have drawn the equator. Every point on that line is the same distance from the north and south poles. These facts may seem simplistic, but they are fundamental to understanding location, direction, distance, time, and the seasons and map projections.

A problem that the ancient Greeks dealt with is locating places on the earth. Location isn't a problem for small areas like a neighborhood. We can describe the grocery as being at the intersection of two streets or a mile to the right from my house and then ½ mile to the left. But how do we describe the location of a place on a sphere 25,000 miles (40,000 km) in circumference? It is like trying to describe where a hole is on a basketball. But unlike the basketball, there are fixed points on the earth that we can use to establish a location system. The poles can be used as starting points. We have already described the equator as a line midway between the poles. The equator is a *great circle*, that is, a circle on a sphere that has the same circumference as the sphere. The equator is the starting line for the coordinate system and has a value of zero. Because we are dealing with spheres and circles, measurements are expressed in degrees, minutes, and seconds, or in decimal degrees. The distance from the equator to the north or south pole is 90° (Figure 4.1). Therefore, we can express the distance

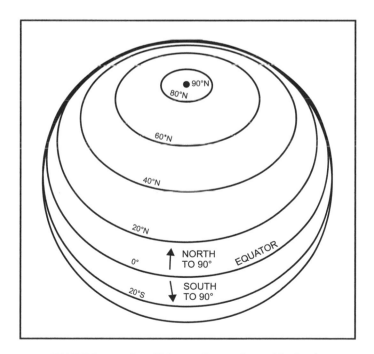

FIGURE 4.1. Parallels are lines of equal latitude.

of any point from the equator by its angular distance (degrees) from the equator. This angular distance is called the *latitude* of the place. We can draw a line around the earth that shows, for example, every place 30° north of the equator. That line is called a *parallel* and is a *small circle*, that is, one that is not the same circumference as the sphere. Note that parallel and latitude are *not* synonymous—parallels are lines and latitude is a distance. This is analogous to a football field. The 50-yard *line* is equidistant from the end zones, but if a player runs from the 20-yard line at one end of the field to the 30-yard line at the other end, he has run a *distance* of 50 yards. The latitude is described by distance in degrees (or angular distance) north or south of the equator.

Most maps now are drawn with north at the top, but on maps of the Middle Ages, east or the Orient was at the top, from which we get the term "to orient the map." We must remember, however, that north is not "up." Up refers to elevation, and north-as-up thinking can lead to serious errors in map reading. An example is assuming that rivers in North America flow "down" to the south and that rivers can't flow north because it is "up." Do not confuse direction with elevation.

Finding the latitude solves only half of our location problem. Since every point on a parallel has the same latitude, we need to determine where along that 30-degree parallel our point lies. The solution is to draw north–south lines that extend from the north pole to the south pole and are half of a great circle (Figure 4.2). These lines are called *meridians*. The grid formed by parallels and meridians is called the earth's *graticule* (Figure 4.3). The intersection of a parallel and a meridian describes the location of a point and is given in terms of latitude and longitude.

Every point along a meridian is the same distance from the starting line, which is called the *prime meridian*. The angular distance east or west of a meridian from the

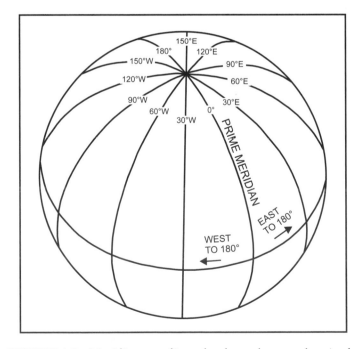

FIGURE 4.2. Meridians are lines that have the same longitude.

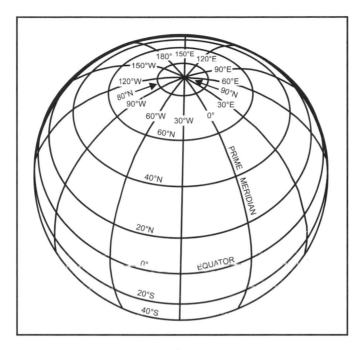

FIGURE 4.3. Earth's graticule.

prime meridian is called the *longitude*. Longitude runs from 0° at the prime meridian, east and west to 180° opposite the prime meridian.

The concept of longitude is simple but was one of the most difficult problems for past geographers to solve. Unlike the poles for latitude, there is no fixed point or line that can be used as a starting line for east–west measurement; most countries used a north–south line through their capital city as a prime meridian. Thus, France used a line through Paris, England used a line through London, and the United States used Philadelphia and then Washington, DC. For a period after the American Revolution, both the London and Philadelphia meridians were shown on some maps of the American colonies. The multitude of meridians used in the world created some confusion if one was using maps created in different countries.

A bigger problem was determining one's longitude, especially at sea. A sailor can determine his/her latitude in the northern hemisphere by measuring the angle between the horizon and Polaris, the North Star. The earth's axis points toward Polaris.[1] Since longitude begins from an arbitrary line, measuring it on a map or globe is easy, but determining longitude in the middle of the ocean on a moving ship could only be done by estimation. This problem resulted in many disasters as ships ran aground or crashed into cliffs with great loss of life. The story of the solution of the longitude is one of the most important and interesting in the history of cartography. Several excellent books on the story of the longitude are listed at the end of the chapter. But to summarize, in 1714 the English Parliament offered a prize of 20,000

---

[1]The earth's axis varies over the course of a 25,000-year cycle so it doesn't point exactly at Polaris, but for purposes of this discussion we can assume it does. This variation is called the "precession of the equinoxes."

pounds to anyone who could find a method of determining longitude at sea. John Harrison, a clockmaker, solved the problem. It was known that one complete rotation of the earth (360°) was one day or 24 hours, and thus, one hour of time would represent a distance of 15° of longitude. Thus, if the navigator knew the difference in time between his starting point, London for example, and the time at his present location, he would know how far he had sailed. The concept is fairly simple, but in the 1770s keeping time on a rolling and bucking sailing ship was impossible. The clocks at the time were pendulum clocks (invented in 1656) that wouldn't function on a ship. Harrison's contribution was the invention of the *chronometer*, a precise timepiece that could be set for London time. The navigator would determine local noon, which is the time the sun is directly overhead, and compare it to London time; the difference would allow him to calculate his longitude. Captain Cook, the famous explorer, carried one of Harrison's chronometers on his second Pacific voyage (1772–1775). The watch on your wrist or the clock on your cell phone is probably more accurate than Harrison's chronometer, but Harrison ultimately, though posthumously, was awarded the prize offered by Parliament.

The Seasons

We noted earlier that the earth's axis points to the North Star. This observation has importance for understanding the seasons. The planets, including the earth, revolve around the sun, on an imaginary plane called the *plane of the ecliptic* in elliptical orbits. The earth's axis is tilted with respect to this plane by 23½°. Because of this tilt, the angle at which the sun's rays strike the earth varies from place to place during the year (Figure 4.4). At the *equinoxes* (about March 21 and September 21), the sun's rays strike the equator at a 90° angle, and they are tangent at the poles (Figure 4.5). At the solstice on June 21, the sun's rays are perpendicular to the parallel at 23½°N and the tangent rays strike 23½° beyond the north pole at 66½°N and no light reaches the south pole. The parallel at 23½°N is called the *Tropic of Cancer*; the parallels at 66½°N and 66½°S are called the *Arctic* and *Antarctic Circles*. On about December 21 the situation is reversed and the direct rays strike 23½°S (*Tropic of Capricorn*) and the area north of the Arctic Circle receives no sun (Figure 4.5). The importance of this tilt is that it causes the seasons on earth; if there were no tilt of the axis, there would be no seasons. The period between September 21 and March 21 in the northern hemisphere is autumn and winter, and in the southern hemisphere it is spring and summer.

OTHER COORDINATE SYSTEMS

The *Universal Transverse Mercator Grid* (UTM) was developed in the 1940s and is another way to locate places on the earth. The rectangular UTM grid divides the earth into 60 north–south zones, each 6° wide in longitude. The zones are numbered from Zone 1, which is between 180° and 174° west longitude, to Zone 60, which is between 174° and 180° east longitude. The conterminous United States is covered by 10 zones as shown in Figure 4.6. Each longitude zone is divided into 20 latitude zones that begin at 80°S and extend to 84°N. Latitude zones are lettered from C in the

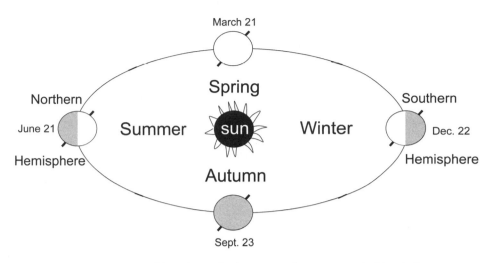

FIGURE 4.4. Earth's axis is tilted 23½° and points to the North Star.

south to X in the north. The letters I and O are not used since they could be confused with numbers and A, B, Y, and Z are used for parts of the Arctic and Antarctic areas.

In each zone, coordinates are measured north and east in meters. *Easting* is the distance of a point in meters from the central meridian of the longitude zone, and *northing* is the distance of a point in meters from the equator. The central meridian through the middle of each 6° zone is given an easting value of 500,000 meters; thus grid values to the west of the central meridian are less than 500,000, and to the east they are more than 500,000 meters. In order to avoid negative numbers south of the equator, the equator is given a northing of 10,000,000 meters. Places are identified by the longitude zone and an *easting* and *northing* coordinate pair (Figure 4.7).

State and local governments widely use the *State Plane Coordinate System* (SPS or SPCS; Figure 4.8), which they consider to be more accurate since the zones are smaller. Like UTM, SPCS is based on zones, but these zones usually follow political

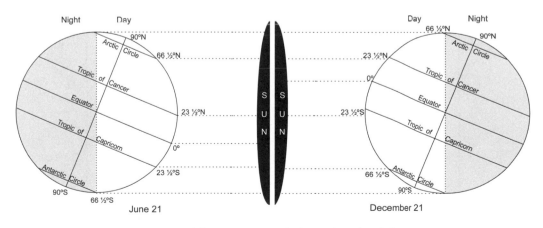

FIGURE 4.5. The seasons result from the tilt of the axis.

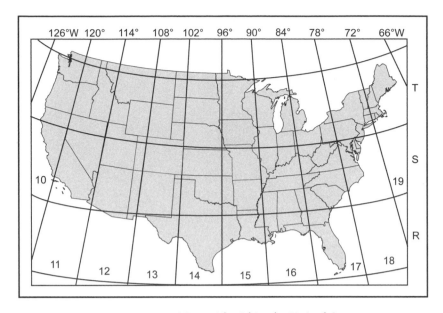

FIGURE 4.6. UTM grid within the United States.

boundaries so the zones are irregular rather than rectangles. The number of zones depends on the size of the state, and, except in Alaska, the zones follow county lines. There are about 120 SPCS zones in the United States, and like UTM, eastings and northings are used; for SPCS they are distances east and west of an origin. The origin is usually 2,000,000 feet west of the zone's central meridian.

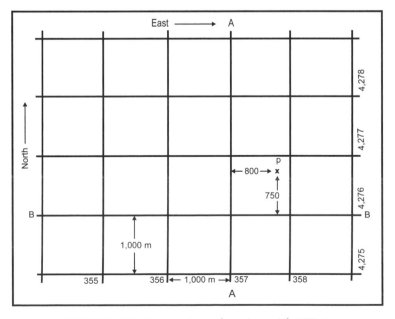

FIGURE 4.7. Determining location with UTM.

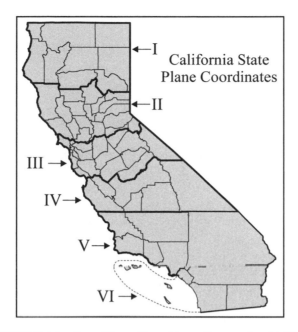

FIGURE 4.8. State Plane Coordinate System for California.

On road maps and some atlas maps an *alphanumeric grid* is often used. The map is divided into a rectangular grid, and the resulting grid has numbers along one border and letters along another. By this means, any place can be located with a letter and number. Usually a place-name index is included on the map (Figure 4.9). The grid does not conform to latitude and longitude.

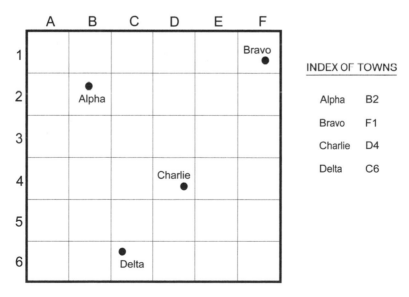

FIGURE 4.9. On some maps an alphanumeric grid is used.

DIRECTION AND DISTANCE

Our globe is magnetic; not in a mechanically uniform way,
but in a temperamentally uneven way.

—*Mapping* (p. 61)

Direction and distance are, in some ways, two of the more difficult concepts in map reading. The simple definition of *direction* is the position of one point on the earth relative to another. The position is measured by an angle between some convenient reference line, such as a meridian, and the line that is the shortest distance between two points. It is here that the difficulty arises. We are accustomed to using the *cardinal directions* of north, south, east, and west in small areas, such as a town, where the distance between the points is apparently a straight line. Because the area is small, the curvature of the earth is not involved; but when we are working at a global scale, the shortest distance between two points is not a straight line but a great circle. If you stretch a string between two points on a globe, it will mark the shortest distance—the most direct line—and it forms a great circle arc. A *great circle* is a circle on the globe or earth that has the same diameter as the globe. A great circle is formed by passing a plane through the center of a sphere (here a globe or the earth), and the line of intersection is a great circle. On Figure 4.10 a great circle is drawn between two points on the same parallel to illustrate this problem. Note that the great circle does not follow the parallel and that the cardinal directions change along the course of the great circle arc.

If we try to follow a line of constant compass direction, that is, one that crosses each meridian at the same angle, the line spirals toward the pole. The line of constant compass direction is called a *loxodrome* or *rhumb line* (Figure 4.11). Only along

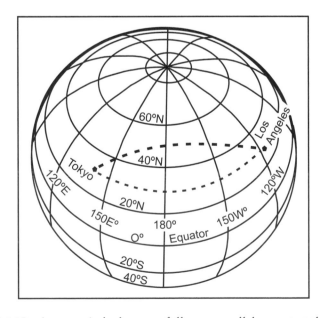

FIGURE 4.10. A great circle does not follow a parallel except at the equator.

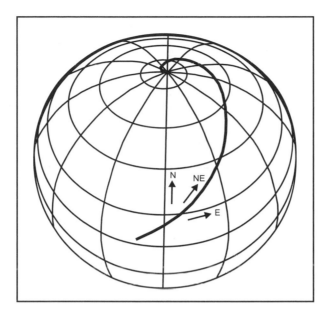

FIGURE 4.11. Loxodrome or line of constant compass direction on the globe.

meridians and the equator, which are great circles, are lines of constant compass direction and cardinal directions the same.

For navigation on the spherical earth, instead of compass directions we use a different type of direction called the *azimuth*. Azimuth is given in degrees and is the angle between a true north–south line (a meridian) and the great circle arc that joins the observer and the point observed. It is measured clockwise from north from 0° to 360° (Figure 4.12). Azimuth applies to small areas as well as the whole earth. In some fields, *compass bearing* is used (Figure 4.13). Bearing is also given in degrees but only 90°, such as N40°E, meaning 40° east of north, or S40°W, meaning 40° west of south.

To add to the confusion, two or three "norths" might be shown on the map

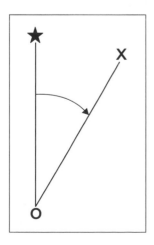

FIGURE 4.12. Azimuth.

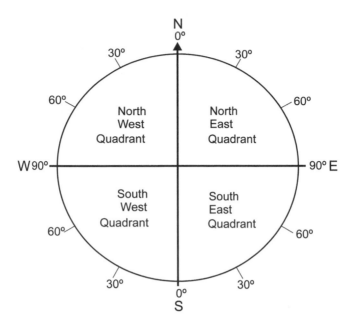

FIGURE 4.13. Compass bearing.

(Figure 4.14). When we usually think of north, we visualize the north pole, and a true north line is a meridian. However, if we are using a compass for navigation, it doesn't necessarily point to the north pole. The compass needle is a magnet, and it points to *magnetic north*. Unfortunately, as Greenhood pointed out, the earth doesn't have a huge magnetic bar on its axis so that the poles coincide with the magnetic poles. Instead, the magnetic north pole is over a thousand miles away from the "true" north pole. On topographic maps (large-scale maps that show the terrain; see Chapter 8) the true north line, the meridian, is shown with a north line topped with a star, but these maps also show a line topped with an arrow that indicates magnetic north, as well as a line marked GN that is *grid north* that points north along the map's grid lines. Both USGS topographic maps and British Ordnance Survey maps show all three lines.

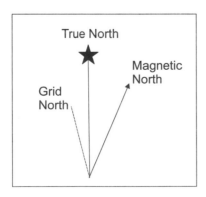

FIGURE 4.14. Grid north, magnetic north, and true north.

CADASTRAL MAPS AND SURVEY SYSTEMS

These "square deal" lines can agree both with the poet
who says, "Something there is that doesn't like a wall," and
with his neighbor who replies, "Good fences make good
neighbors."

—*Mapping* (p. 23)

Cadastral maps are maps of property ownership and are one of the oldest types of
map. Any society that recognizes property ownership has need for such maps. These
maps are also often used as locational systems. Cadastral maps are based on surveys
of the land that may be either unsystematic (i.e., irregular) or systematic (i.e., regular).
In North America, several different survey types, both unsystematic and systematic,
have been used, and these surveys have left their marks on the land. When we fly over
the continent and look down, we can see patterns of roads and fields, wood lots, and
grazing lands; these patterns are largely a result of cadastral survey systems.

Irregular Systems

In Europe and early English-settled parts of the United States and Canada, *metes and
bounds* was the first method used. In metes and bounds the settler purchased or was
given a certain amount of land in an area; he then selected the parcel he wanted and
surveyed it or had it surveyed. At this time surveying was a common subject in a boy's
education. The settler would usually take the best land available, free of swamps and
other undesirable land. The property description was based on measurements from
various points, usually natural features. A description might begin: from the willow
tree at the river's edge N32°E 400 yards to the large boulder, thence E 150 yards to
the oak tree and S to the river and along the river back to the starting point. The
survey measured (meted) and bounded the property (Figure 4.15). The result of metes

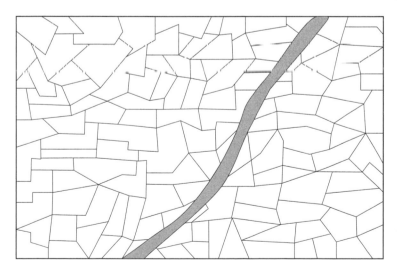

FIGURE 4.15. Metes and bounds property patterns.

and bounds is a crazy-quilt pattern of land holdings, and since the description of properties depends on features that are not permanent, quarrels were common. Most such areas have now been re-surveyed by more precise methods; permanent markers have been placed at corners, and the boundary lines have been established and recorded; this method has eliminated hostilities. Since roads often follow boundary lines, the road patterns in metes and bounds areas are also irregular.

French-settled areas, especially Louisiana and along the St. Lawrence River, divided the land by a method called the *seigneurial* system, frequently called the *long lot* system or *French Long Lots*. In this system, soldiers or other chosen citizens were granted a parcel of land along a waterway. In order that each parcel or *arpent* would have water frontage, the lots were long and narrow. Thus, these long lots had space for a dock plus elevated land for a house and other buildings and rich farm land. Long lots result in a very distinctive pattern (Figure 4.16). These arpents are now given numbers similar to section numbers (see below).

Parts of the United States were also settled by Spanish or Mexican colonists. Spain and later Mexico provided land grants that were often quite large to settlers, either as individuals or as groups. These grants were given in Florida, Texas, Arizona, New Mexico, and California, and they were surveyed in a metes and bounds manner. These grants were the basis of the ranchos of California; rancho maps were called *diseños*. Evidence of the old survey patterns can be seen in street patterns in Los Angeles and other Southern California locations.

A somewhat more systematic arrangement, though still having irregular boundaries, was the *New England Town* or *township*. These towns were roughly 6 miles apart and 36 miles square and were *civil townships*; that is, they functioned in much the way a county does today.

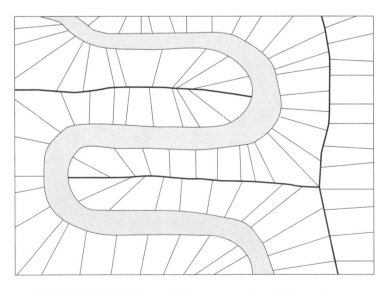

FIGURE 4.16. Seigneurial system or French Long Lots.

Systematic Surveys

In 1785, the United States Congress enacted the Land Ordinance, which provided for systematic survey *before* the land was settled. This was the beginning of the *United States Public Land Survey System*, usually called PLSS and sometimes the *rectangular survey system*. The system was the brainchild of Thomas Jefferson and was established under the Northwest Ordinance of 1787. This ordinance established a rectangular survey system that was designed to aid in the transfer of federal lands to private citizens. It was not the first systematic survey—indeed the Romans used a regular system—but PLSS was unique in that it provided for survey *before* settlement and was more equitable than metes and bounds. Over 1.5 billion acres (640 acres is 1 square mile) have been surveyed under PLSS. Nineteen states are not covered by PLSS. They are the existing states at the time of its establishment (West Virginia and Kentucky were part of Virginia, Maine was part of Massachusetts) and Texas, which had its own survey system when it joined the Union. Figure 4.17 shows the covered states.

PLSS began in the state of Ohio, which was somewhat of an experimental area. Figure 4.18 shows the variety of systems used in the state. The beginnings of the rectangular system was in the "Seven Ranges," but in other areas, especially the northeast and south, the metes and bounds system was used. The northwest corner of Ohio marks the beginning of the final system.

PLSS is a very logical and easy-to-understand system that is based on a series of named *principal meridians* and numbered *base lines* (Table 4.1). There are 37 sets of these meridians and base lines in the United States. The principal meridians actually make up a set of separate surveys. Extending east and west of a given principal

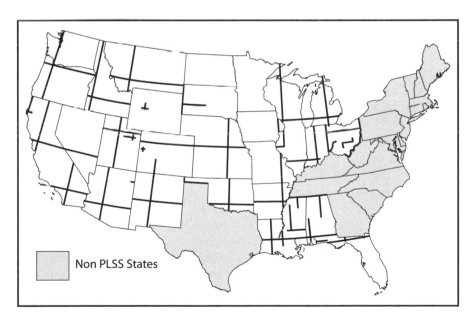

FIGURE 4.17. States covered by PLSS and principal meridians.

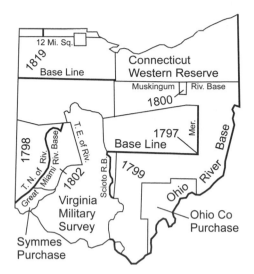

FIGURE 4.18. Ohio was an experimental area for survey systems.

meridian and spaced at 24-mile intervals north and south of a base line are *standard parallels* (often called *correction lines*). *Guide meridians* extend north from the base line or standard parallels at intervals of 24 miles east and west of the principal meridian (see Figure 4.19). The correction lines compensate for the convergence of meridians toward the poles, thus, the northern border of these tracts is slightly less than 24 miles. Because the road network commonly follows the township and range lines, one can find roads that appear to be offset, either by sharp corners or smoothed into a curve with no physical feature to account for it. This occurred because the road planners following the range line exactly included the "jog" at the correction line.

The parcels of land enclosed by the standard parallels and guide meridians are further subdivided into *townships*. A township is 6 miles on a side or 36 square miles (23,040 acres). The lines forming these townships are *tiers* and *ranges*. The townships are designated by their tier and range number, such as Tier (or township) 22 north, Range 2 east. Townships are further divided into *sections*, each one square mile or 640 acres in area. Sections are numbered from 1 to 36, with "1" in the northeast corner and proceeding alternately east and west to "36." Sections, too, can be subdivided into ½ sections, ¼ sections, and smaller units; these subdivisions are also called *aliquot parts*. Sometimes these small subdivisions are further divided into numbered *government lots* that are not necessarily rectangular.

This system allows for any plot of land to have a unique designation. For example, Mount Diablo Meridian, Nevada, T. 19 S., R. 60 E., Sec. 32, SE1/4 NW1/4NW1/4NW1/4, is the legal description of a 2½ acre piece of land in Nevada located in section 32 of the 19th tier south of the base line, the 60th range east of the Mount Diablo principal meridian in the southeast quarter *of* the northwest quarter *of* the northwest quarter *of* the northwest quarter. Notice that the description is read "backwards"; commas are not included in the official description but may be included in other kinds of text for ease of reading (Figure 4.20).

Sometimes one finds areas where the sections are not perfectly square or the lines

TABLE 4.1. Principal Meridians

Meridian	Latitude/Longitude of origin	Governing surveys in the states of . . .
Black Hills	43° 59' 44"N; 104° 03' 16"W	South Dakota
Boise	43 22 21; 116 23 35	Idaho
Chicasaw	35 01 58; 89 14 47	Mississippi
Choctaw	31 52 32; 90 14 41	Mississippi
Cimarron	36 30 05; 103 00 07	Oklahoma
Copper River	61 49 04; 145 18 37	Alaska
Fairbanks	64 51 50.048; 147 38 25.949	Alaska
Fifth Principal	34 38 45; 91 03 07	Arkansas, Iowa, Minnesota, Missouri, North Dakota, South Dakota
First Principal	40 59 22; 84 48 11	Ohio and Indiana
Fourth Principal	40 00 50; 90 27 11	Illinois
Fourth Principal (extended)	42 30 27; 90 25 37	Minnesota and Wisconsin
Gila and Salt River	33 22 38; 112 18 19	Arizona
Humboldt	40 25 02; 124 07 10	California
Huntsville	34 59 27; 86 24 16	Alabama and Mississippi
Indian	34 29 32; 97 14 49	Oklahoma
Kateel River	65 26 16.375; 158 45 31.014	Alaska
Louisiana	31 00 31; 92 24 55	Louisiana
Michigan	42 25 28; 84 21 53	Michigan and Ohio
Mount Diablo	37 52 54; 121 54 47	California and Nevada
Navajo	35 44 56; 108 31 59	Arizona
New Mexico Principal	34 15 35; 106 53 12	Colorado and New Mexico
Principal (of Montana)	45 47 13; 111 39 33	Montana
Salt Lake	40 46 11; 111 53 27	Utah
San Bernardino	30 07 13; 116 55 48	California
Second Principal	38 28 14; 86 27 21	Illinois and Indiana
Seward	60 07 37; 149 21 26	Alaska
Sixth Principal	40 00 07; 97 22 08	Colorado, Kansas, Nebraska, South Dakota, Wyoming
St. Helena	30 59 56; 91 09 36	Louisiana
St. Stephens	30 59 51.463; 88.01 21.076	Alabama and Mississippi
Tallahassee	30 26 03; 84 16 38	Florida and Alabama
Third Principal	38 28 27; 89 08 54	Illinois
Uintah	40 25 59; 109 56 06	Utah
Umiat	69 23 29.654; 152 00 04.551	Alaska
Ute	39 06 23; 108 31 59	Colorado
Washington	30 59 56; 91 09 36	Mississippi
Willamette	45 31 11; 122 44 34	Oregon and Washington
Wind River	48 00 41; 108 48 49	Wyoming

aren't straight; these usually arise from errors in early measurements and are often found in hilly areas.

In Canada, a similar system is in use for most of western Canada called the *Dominion Land Survey* (DLS) that was begun in 1871. Like PLSS, on which it is modeled, DLS is based on a series of numbered meridians and base lines. The first meridian is just west of Winnipeg, Manitoba, and the others are spaced at four-degree intervals to the coast meridian. Townships are 6 miles (9.7 kilometers) on a side, and like those in the United States are designated by their township and range number. On maps, township numbers are marked in Arabic numerals, and range

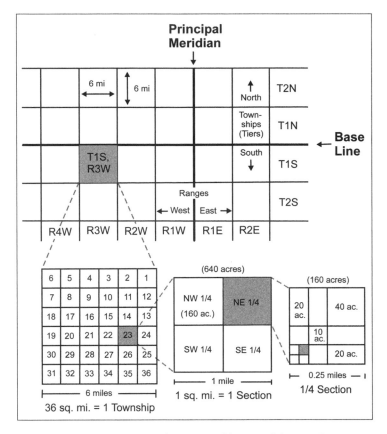

FIGURE 4.19. United States Public Land Survey System.

numbers are sometimes in Roman numerals and sometimes in Arabic. Like PLSS, townships are divided into 36 sections, but unlike in PLSS the numbering starts with "36" in the northeast corner alternating to "1" in the southeast corner.

Maps that specifically show property are called *plat maps* or *cadastral maps* and are official records of land ownership. Topographic maps do not include individual properties, but the survey system is shown, and thus for our Nevada example above, we can plot the parcel of land and learn much about the nature of its surroundings.

FURTHER READING

Harris, Lucia Carolyn. (1960). *Sun, Earth, Time, and Man.* Chicago: Rand McNally.

Hubbard, Bill, Jr. (2009). *American Boundaries: The Nation, the States, the Rectangular Survey.* Chicago: University of Chicago Press.

Johnson, Hildegard Binder. (1976). *Order upon the Land.* New York: Oxford University Press.

Kain, Roger J. P., and Elizabeth Baigent. (1992). *The Cadastral Map in the Service of the State: A History of Property Mapping.* Chicago: University of Chicago Press.

Nevada 026060

The United States of America,

To all whom these presents shall come, Greeting:

WHEREAS, a Certificate of the Land Office at   Reno, Nevada,
is now deposited in the Bureau of Land Management, whereby it appears that full payment has been made
by the claimant s   Francis X. Zink and Virginia R. Zink,

pursuant to the provisions of the Act of Congress approved June 1, 1938 (52 Stat. 609), entitled "An
Act to provide for the purchase of public lands for home and other sites," and the acts supplemental there-
to, for the following-described land:

Mount Diablo Meridian, Nevada.

T. 19 S., R. 60 E.,

Sec. 32, SE¼NW¼NW¼NW¼.

The area described contains 2.50 acres, according to the Official Plat of the Survey of the said Land,
on file in the Bureau of Land Management:

FIGURE 4.20. Official property description for a 2.5-acre parcel of land in Nevada.

Linklater, Andro. (2002). *Measuring America: How an Untamed Wilderness Shaped the United States and Fulfilled the Promise of Democracy.* New York: Walker.

Price, Edward T. (1995). *Dividing the Land: Early American Beginnings of Our Private Property Mosaic.* Chicago: University of Chicago Press.

Sobel, Dava. (1995). *Longitude: The True Story of a Lone Genius Who Solved the Greatest Scientific Problem of His Time.* New York: Walker.

Thrower, Norman J. W. (1966). *Original Survey and Land Subdivision: A Comparative Study of the Form and Effect of Contrasting Cadastral Surveys.* Chicago: AAG and Rand McNally.

White, C. Albert. (1991). *History of the Rectangular Survey System.* Washington, DC: U.S. Government Printing Office.

RESOURCES

Cadastral Slide Show
   www.blm.gov/cadastral/cadcom/cadmedia.htm
Public Land Survey System
   www.nationalatlas.gov/articles/boundaries/a_plss.html
UTM Fact Sheet
   pubs.usgs.gov/fs/2002/0077/report.pdf

CHAPTER 5

Map Projections: The Round Earth on Flat Paper

> Just as a building's foundation must come to terms
> with the ground, a map's projection must come to
> terms with the earth.
>
> —*Mapping* (p. 114)

A map's accuracy depends on many factors, such as its scale, the amount of generalization, and the mapmaker's skills, but when it comes to representing the entire earth, *all* maps are inaccurate to some degree. Two solid figures, cones and cylinders, are called *developable figures* because they can be cut apart and flattened without any distortion. The sphere is not a developable figure. The earth (or a globe) cannot be transformed into a flat map exactly. There will be stretching, tearing, compression, or all three, somewhere on the map. Only the globe can represent the earth's shape exactly.

If flat maps do not represent the earth accurately, why do we use them? Why not just use a globe? As it happens, globes also have limitations. The first of these limitations is scale. While a map might have a scale of 1:1,000,000, a globe of this size would be over 500 miles (800 km) in diameter! It is an understatement to say that a globe this size would be impossible. Second, consider the number of maps that you find in an average atlas; even if you only concentrate on the world maps, over 100 globes might be needed to represent the different topics. Librarians would not have enough room on their shelves to house globes showing the information of a single atlas that would take only 1 inch of shelf space. Even small globes, like the 4- to

6-inch (10–15 cm) "pocket globes" that were popular in the 18th century, are cumbersome to carry around. Finally, we can only see one-half of the globe at a time and, for many purposes, we would like to see the whole world at a glance. What about virtual globes, such as Google Earth, then? Here again there are some problems. While these are wonderful tools and increasingly available on portable devices, they do not yet have all of the features of the many kinds of flat maps available, including the whole earth at a glance.

When cartographers create a flat map of the earth, they do it in a systematic manner; they do not simply flatten a globe by cutting or stretching. The method of creating a flat map is called *projection*. The formal definition of *map projection* is *a systematic representation of the earth's grid upon a plane,* which simply means putting the round earth on flat paper.

At this point you might ask, "Why is this important in map *reading*? Why do I need to understand map projections?" To begin with, different projections have different properties; for example, one might show shapes correctly, but not areas; another projection might show distances correctly, but not directions, or a map might be distorted at the edges, but accurate in the center. It is only by understanding these properties and distortions that you can get the most and best information from a map and not be misled by distortions. The following discussion is a nonmathematical, basic explanation of map projections. An infinite number of projections are possible, but we can only look at a few of the most common ones here; for those who want to explore projections in more depth, or with a more mathematical treatment, several references are listed at the end of the chapter. The map projection poster available from USGS is an especially useful tool. (See Resources at the end of the chapter.)

PROJECTION PROPERTIES

Certain properties of the earth, when represented on a globe, are important to review before looking at map projections. Any distortion of these properties means there is deformation on the map:

1. All parallels are parallel.
2. Meridians are half great circles that converge at the poles.
3. Meridians are evenly spaced along any parallel.
4. "Rectangles" formed by the same two parallels and having the same longitudinal dimensions will have the same areas.
5. Parallels are equally spaced along the meridians (assuming the earth to be a sphere).
6. Area scale is uniform; that is, 1 square inch of the globe in the high latitudes and 1 square inch in the low latitudes cover the same earth area.
7. Distance scale is uniform.

Map projections are placed in categories based on the properties they preserve. Thus, there are *equal-area* projections that show sizes correctly, *conformal* projections that show very small shapes correctly, *azimuthal* projections that show azimuths correctly, and *equidistant* projections that show distances correctly. *All of these cannot be found on the same map.* Especially important: *there is no such thing as an equal-area, conformal projection*; the two properties are mutually exclusive, as we will see below.

Equal-area projections show sizes of areas correctly at the expense of shapes. In order to achieve equivalence of area, if there is stretching in one direction, there must be compression in the orthogonal (at right angles) direction. Figure 5.1 shows several shapes that all have the same area. This is the only property that can be achieved over the entire map.

Conformal projections require that parallels and meridians cross at right angles and that the scale be the same in all directions about a point. In simple terms, stretching in latitude is met with stretching in the same amount in longitude, and compression in latitude is matched with compression in longitude. For an illustration, see the shapes in Figure 5.2. As a result, angles are correct, but areas are severely distorted. On a conformal map, square areas that are 100 square miles or 100 square kilometers in different parts of the globe will remain square, but their sizes will vary, often greatly. It is a mistake to call conformal maps "true shape." Conformality is actually only true for very small shapes like a bay or small island; however, the overall shapes, while distorted, are recognizable (see Figure 5.3). Figure 5.4 shows Greenland on both a conformal and an equal-area projection. Because of the ways in which the two types of projection are created, we can see that the two properties cannot occur on the same map.

Equidistant projections show distances correctly, but *only along certain lines or from specific points*; equivalence cannot be achieved over the entire map. *Azimuthal projections*, of course, show azimuths correctly. Recall from Chapter 4 that azimuth is a specific way of expressing direction. An azimuthal projection does not show the cardinal directions (N, S, E, W) correctly, and calling these projections true direction can be misleading.

Finally, some projections are compromises and are sometimes called *minimum error projections* or *compromise projections*. These projections have no particular properties except appearance; they look good, and as a result are frequently used when determining distance, size, direction, and shape is not important. While they

FIGURE 5.1. These figures all have the same area.

FIGURE 5.2. The two columns of figures have the same shapes, but their areas differ.

aren't accurate for any of these properties, they "aren't too bad" in any of them since there are no major distortions.

PROJECTION TYPES

As we learned earlier in this section, cartographers create projections in a systematic and orderly manner. But what does this mean? We also learned that there are "developable surfaces" that can be flattened into a plane without distortion or tearing. The cartographer uses these surfaces by putting them in contact with a hypothetical *generating globe* and "projecting" the grid onto the surface. The projection can be done by using a transparent globe and a light source (for demonstration purposes only), by geometric projection, or, in the actual way cartographers create them, by mathematical tables and formulas.

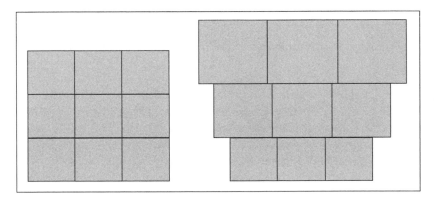

FIGURE 5.3. Conformal maps show shapes correctly for small areas, but the overall form is distorted.

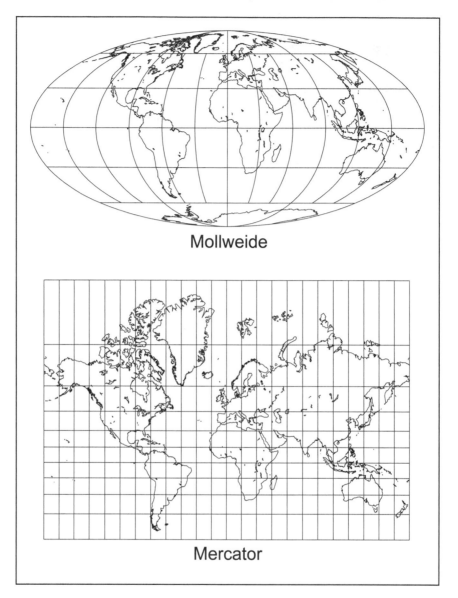

FIGURE 5.4. Comparison of equal-area (homolosine) and conformal (Mercator) projections.

All projections have a *zone of best representation*, also called the *area of least deformation* (Figure 5.5). Within that zone, there is little error in any of the properties, but the zones are specific for each projection. If you are familiar with the zone of best representation on a projection, you know where there is the least amount of error for distance, size, and angles.

Projections can be categorized by the way in which they were hypothetically created; thus we have cylindrical projections, conic projections, and plane projections. Finally, we also have a group that cannot be projected onto a developable surface even in imagination; these are called *mathematical* or "*conventional*" projections.

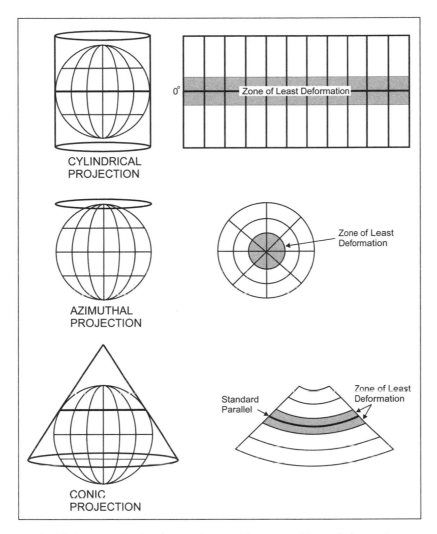

FIGURE 5.5. Projection surfaces with areas of least deformation.

Cylindrical Projections

If we wrap a cylinder around a globe, and project the parallels and meridians onto it, as in Figure 5.5, the result is a *simple cylindrical projection* (sometimes called a *rectangular* projection). As you can see, (1) the parallels and meridians are straight lines and cross at right angles, (2) the spacing of the parallels increases as they approach the poles, and (3) because they are straight lines, the meridians do not converge at the poles. The poles themselves cannot be shown. The zone of best representation is a short distance away from either side of the line where the cylinder touches the globe, the equator in this case. Although the simple cylindrical projection can be projected graphically, it has no special properties; however, if it is "tweaked" mathematically, we can have conformal and equal-area cylindrical projections.

The most familiar cylindrical projection is the *Mercator* (Figure 5.6), which was

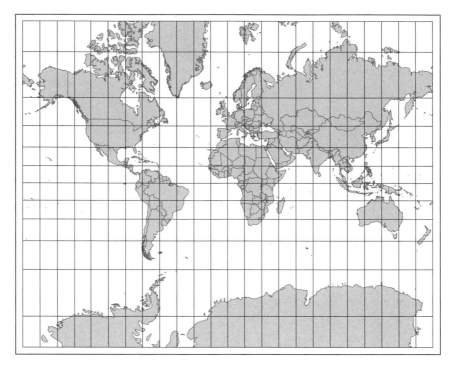

FIGURE 5.6. The Mercator projection.

devised in 1569 by Gerard Mercator for ocean navigation. Because the meridians do not converge at the poles, there is increasing stretching the farther away from the equator one goes, which results in increasing E–W distortion; since the parallels get farther apart as you move toward the poles, there is also great N–S stretching. The projection, like other cylindricals, is most accurate along the equator. The Mercator projection has two major properties that make it useful for navigators: (1) it is conformal and (2) *rhumb lines* are shown as straight lines on the map. Since it is conformal, the shapes, especially of small areas, are correct, but more important for Mercator, and his reason for creating the projection, are the straight rhumb lines. A rhumb line or *loxodromic* curve is a line of constant compass direction. As we saw earlier, if we follow a NE line (45° from N) on a globe, the line spirals toward the north pole. But on the Mercator projection, a line of constant compass direction between two points is straight, making it comparatively easy to sail between the points with only a compass (Figure 5.7). The distance is not the shortest distance between the two points; that would be a great circle, but a great circle course would require multiple shifts in compass heading. Only the Mercator projection has this property; none of the other cylindrical projections do, nor, in fact, do any other projections.

The Mercator projection has been much misused. Sea navigation is still its best use, but because of its tidy rectangular shape and ease of drawing, it was a standard map for schoolrooms and textbooks from the early 19th century until World War II when air navigation became common. Until that time, schoolchildren wondered why Greenland was an island and Australia a continent, since Greenland was so much bigger on the map. Actually, Greenland is the size of Mexico.

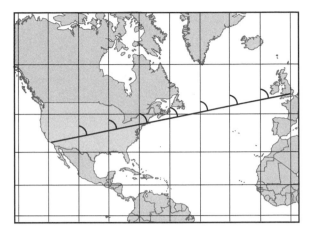

FIGURE 5.7. A rhumb line (line of constant compass direction) on a Mercator projection.

By adjusting the spacing of the parallels on the cylindrical, a *cylindrical equal-area projection* can be created (Figure 5.8). Because the meridians are straight parallel lines, in order for the projection to be equal-area, the parallels must get closer together as they approach the poles, thereby matching stretching N–S with compression E–W.

There are other cylindrical projections, such as the *Miller cylindrical* (see Appendix E), that are neither equal-area nor conformal, but a compromise that has a better appearance than the Mercator. One cylindrical projection that has generated much controversy in the past 40 years is the so-called *Peters projection* (Figure 5.9). This projection was promoted by Arno Peters as a replacement for the Mercator projection. Peters contended that cartographers were biased toward developed countries that are largely in the midlatitudes and used the Mercator projection because those countries were shown larger than underdeveloped countries in the equatorial regions. Peters claimed that his projection, which was actually developed by James Gall in 1851, showed shapes and areas better than the Mercator. As a result, many organizations adopted what became known as the Gall–Peters projection for atlases and wall maps. However, this projection has a number of limitations, some of which are built into any cylindrical projection, and it is highly debatable that the shapes of continents

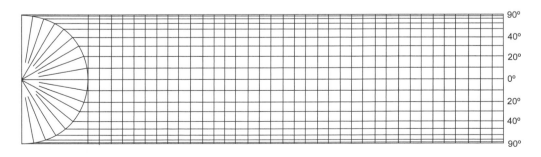

FIGURE 5.8. Cylindrical equal-area projection.

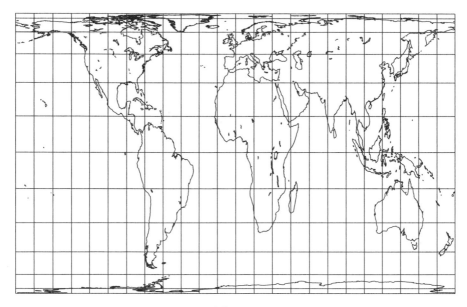

FIGURE 5.9. The Peters projection.

are better. Thus, in 1989, the American Cartographic Association passed a resolution that strongly urged publishers, governmental agencies, and the media to stop using *any* rectangular world maps for general purposes and artistic displays. Unfortunately, despite this resolution, rectangular maps are still found in many children's geography workbooks.

Other Aspects of Cylindrical Projections

For all of these examples, we wrapped the cylinder around the equator of the globe, but we can also wrap it around the poles or, in fact, any great circle. This gives us three *aspects*: equatorial, transverse, and oblique (Figure 5.10). Thus, there are transverse and oblique versions of the Mercator, neither of which shows rhumb lines as straight lines. Probably the most common is the *transverse Mercator projection* (Figure 5.11). This projection is used as the basis for USGS topographic maps and for many State Plane Coordinate Systems. The cylinder is wrapped around a meridian, giving a zone of least deformation about 15° to each side, and it also retains the conformal property of the Mercator projection in the equatorial case. This makes it ideal for topographic maps, but it is only seen for the entire earth for illustrative purposes as in Figure 5.12.

Conic Projections

Just as we can wrap a cylinder around a globe and "project" the grid, we can place a cone on the globe and project the parallels and meridians. When the cone is flattened, the resulting simple conic projection has radiating meridians and parallels that

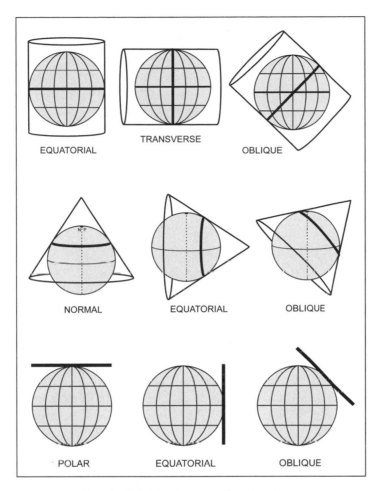

FIGURE 5.10. Aspects of projections.

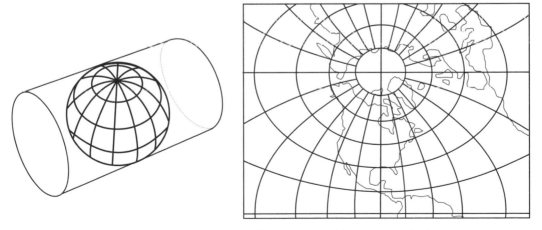

FIGURE 5.11. Formation of the transverse Mercator projection.

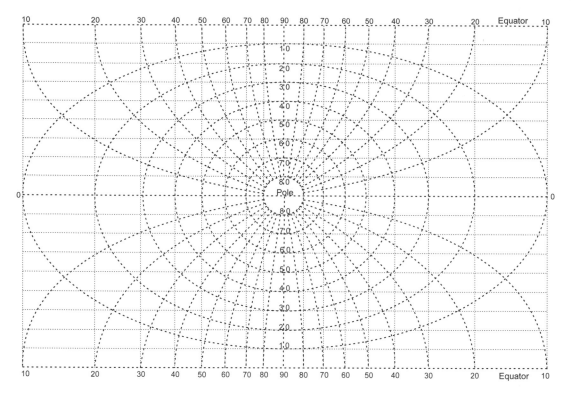

FIGURE 5.12. Transverse Mercator projection for the earth (the dotted lines represent the conventional Mercator).

are concentric arcs. The area of best representation is along the parallel that touches the globe, called the standard parallel. Like the cylindricals, conic projections may be conformal, equal-area, or compromise.

Conic projections cannot show the entire earth and are most often used to represent countries in the midlatitudes, such as the United States or Australia, which are in the zone of best representation.

It is possible to create a conic that has two standard parallels by using a *secant cone*, that is, one that cuts the globe. By choosing the parallels where the cone intersects the globe carefully, the two zones of best representation are brought close together to create a wide area of little deformation. The two most common projections for representing the United States are secant conics: *Lambert's conic conformal* and *Albers' conic equal-area*. When these projections are used for the United States, they have only a small amount of distortion (Figure 5.13). For the conformal, areal distortion of the United States is 5% at a maximum and the linear scale error (distance) is only 2.5%. The equal-area represents shapes quite well, and the maximum linear error is only 1.25%.

In these examples, the cone is placed on a parallel, but like cylindrical projections, we can create oblique and transverse aspect conic projections, as well as the secant conics above.

Plane (Azimuthal) Projections

The globe can also be projected onto a flat surface, and the resulting projections are the *plane projections*. Because all of these projections show azimuths correctly from the center, they are commonly called *azimuthal projections*. Occasionally, the term *zenithal projection* is used. If the plane is tangent at one of the globe's poles, it is in the *polar aspect*; at the equator, it is in the *equatorial aspect*; and at any other point on the globe, it is in the *oblique aspect*. There are five commonly used plane projections, three of which can be created by geometric projection, and all can be used in any of the aspects or cases. On all of them straight lines through the center are great circles; the five differ in the spacing of the parallels (Figure 5.14). Because the plane touches the generating globe in the center of the projection, the center is the area of best representation. Three of these projections can be created with a wire globe and a light source.

If a light source is assumed to be at infinity, the light rays appear to be parallel; if we project the globe this way, the resulting azimuthal projection is called the orthographic projection. The orthographic projection cannot show more than one hemisphere at a time and has no special properties other than azimuthality. The parallels get closer together as you go away from the center. Although, as with any azimuthal projection, it can be centered on the poles, at any point on the equator, or anywhere else on the earth, it is often seen in the equatorial aspect because that view resembles what we see when looking at a globe. Beginning in the 17th century and

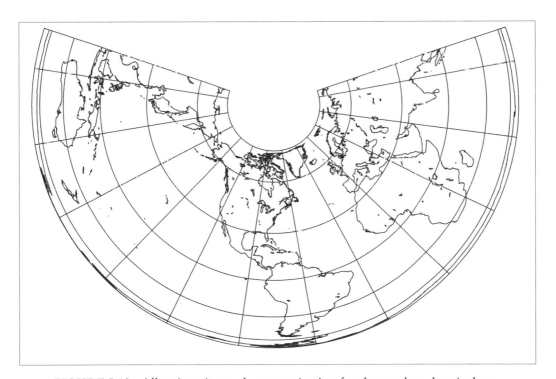

FIGURE 5.13. Albers' conic equal-area projection for the northern hemisphere.

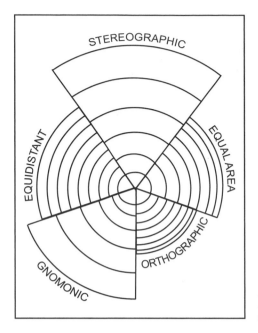

FIGURE 5.14. Spacing of the parallels for the commonly used plane (azimuthal) projections.

continuing until the Age of Space Flight, it was the most common projection for maps of the moon, because when we look at the moon, it appears to be on the orthographic projection (Figure 5.15).

The *stereographic projection* is theoretically constructed by a light source opposite the center of the projection. Thus, if the projection is centered on the north pole, the light source is assumed to be at the south pole. It is possible to show more than one hemisphere but less than the entire sphere; usually only a hemisphere is shown, to represent the entire earth; two hemispheres (usually eastern and western) are used side by side (Figure 5.16). A pair of stereographic hemispheres was commonly used to represent the world in geography books and atlases until the 19th century. Even Mercator used this arrangement in his atlas of the 16th century to represent the world; his cylindrical projection was only used for navigational charts. On the stereographic projection in the polar case, the parallels get farther apart as you go out from the center in the same proportion as the distance between meridians. Thus, this is a conformal projection.

If the light source is imagined to be in the center of the globe, the *gnomonic projection* is produced (Figure 5.17). On the gnomonic (pronounced *no mon′ ic*), it is impossible to show an entire hemisphere, and shapes and areas are highly distorted. Nonetheless, it is a widely used projection because it has one important quality: All great circles appear as straight lines, and all straight lines are great circle paths. This property, combined with azimuthality, makes it important to navigation because great circle routes can easily be plotted to determine the shortest distance between points. It can then be paired with the Mercator projection to plot courses. The gnomonic is used to determine the great circle route, and the Mercator is used to plot short constant compass segments to approximate the great circle. (See Figure 5.18.) Of course, in actual navigation, software helps with the calculations.

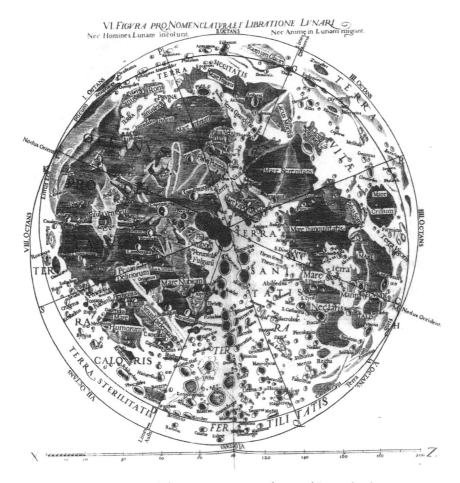

FIGURE 5.15. The moon on an orthographic projection.

Two other azimuthal projections that are in common use are the *azimuthal equalarea* and the *azimuthal equidistant* projections. Neither of these projections can be created with an imaginary light source, but they can be constructed graphically. The equal-area has parallels that get closer together the farther you get from the center, and almost the entire earth can be shown. Shapes are progressively more distorted the farther away they are from the center, but areas can be compared. The azimuthal equidistant is often found in airports and centered on that point (Greenhood called this the "hometown-centered map"). Because the distance shown from the center point is correct, airport visitors can determine how far it is to any other place in the world. This only works from the center, however; distances between other places will not be correct (Figure 5.19).

Mathematical Projections

The mathematical, or conventional, projections include those that cannot be imagined as being created with a transparent globe and light source or by geometric pro-

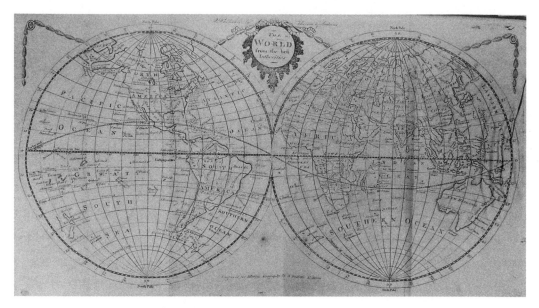

FIGURE 5.16. The stereographic projection showing two hemispheres (from Jedidiah Morse, *Universal Geography*, 1797).

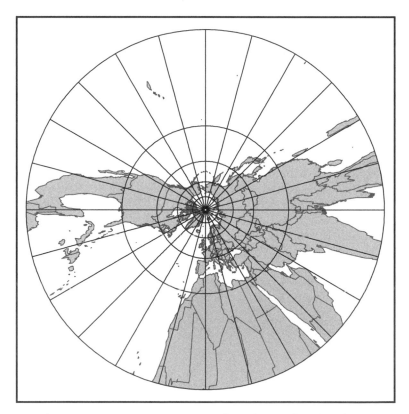

FIGURE 5.17. The gnomonic projection centered on the north pole. This projection cannot show an entire hemisphere.

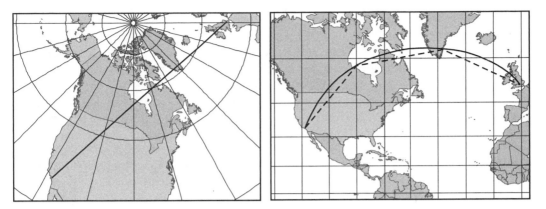

FIGURE 5.18. The gnomonic and the Mercator can be used together in navigating a great circle route. The great circle route is plotted on the gnomonic, and short constant-compass legs are plotted on the Mercator.

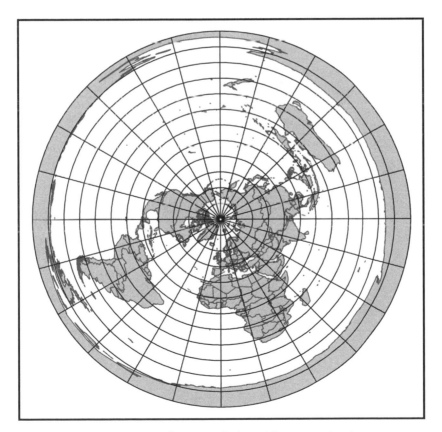

FIGURE 5.19. The azimuthal equidistant projection.

jection. Their shapes vary widely from oval to heart-shaped and star-shaped and even armadillo-shaped; some look like the earth has been peeled like an orange. From the hundreds of such projections that have been devised, we will only look at four that have been or are, commonly used, especially for world maps.

The *sinusoidal* projection is the oldest of the four and was apparently first used by Nicholas Sanson in about 1650; it is also called the Sanson–Flamsteed projection. It is an equal-area projection that has a straight central meridian that is true to scale and equally spaced, true-to-scale parallels. The meridians are trigonometric sine curves, and the projection can be centered on any meridian. Thus, North America or any other location can be in the center of the map. The zone of best representation is around the central meridian and the equator. Shapes become very distorted in the high latitudes, especially at the edges of the map (Figure 5.20).

The *Mollweide*, also called the *homolographic*, projection, devised in the 19th century, resembles the sinusoidal in some ways. It is equal area and can show the entire earth; it has a straight central meridian and straight parallels. However, it is an ellipse, and only the 40th parallels north and south are correct in length. The parallels are not spaced truly on the central meridian. Shapes on the Mollweide are less distorted in the polar regions than on the sinusoidal. The two zones of best representation are found around the central meridian and the 40th parallels. Like the sinusoidal, it can be centered along any meridian (Figure 5.21).

The third of our mathematical projections is the *Goode's homolosine projection*. As the name suggests, it is a combination of the "best" parts of the homolographic and sinusoidal projections. It is made up of the sinusoidal from 40°N to 40°S, and the homolographic from 40° to the poles; thus the zone of good representation is extended. Like its "parent" projections it is equal area, and like the homolographic it has better shapes in the polar regions. An easy way to recognize this projection is by

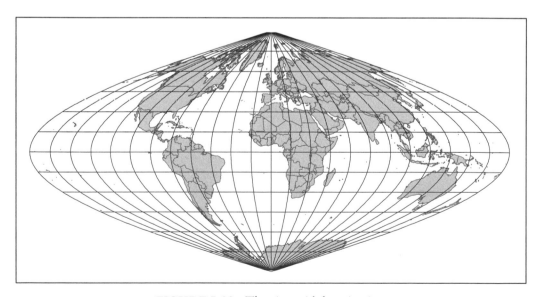

FIGURE 5.20. The sinusoidal projection.

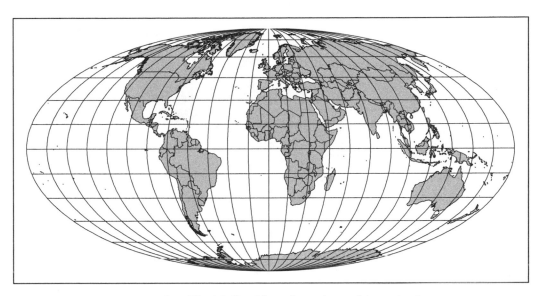

FIGURE 5.21. The Mollweide or homolographic projection.

the "kinks" in the meridians at 40°N and S. J. Paul Goode devised the projection in 1923, and it was used extensively in *Goode's World Atlas* (Figure 5.22).

In 1963, Arthur H. Robinson invented a new projection, which he named the *orthophanic* projection, meaning "right appearing." The name didn't catch on, but the projection did; it has been called the *Robinson projection* almost from the beginning. As Robinson's name for it implies, it looks good; it is a compromise projection—it is not equal-area, it is not conformal, it is not equidistant, and azimuths are not shown correctly from any point. While it does not eliminate distortion, the distortions are relatively small over the map. It has been widely adopted for maps of the entire world in atlases, books, and wall maps (Figure 5.23).

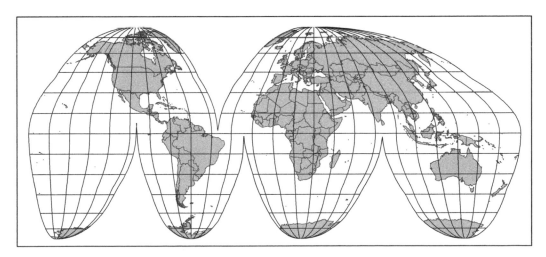

FIGURE 5.22. Goode's homolosine projection, interrupted.

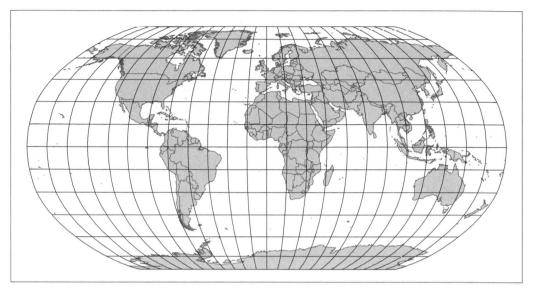

FIGURE 5.23. The Robinson projection.

Interrupted and Condensed Projections

With projections like the Mollweide, the sinusoidal, and the homolosine, it is possible to interrupt the graticule and create more zones of good representation. The cartographer draws a line for the equator and marks it in degrees; then a central meridian is chosen for each continent, and the projection graticule is constructed for that segment of the world. The result is a series of lobes, each made up of a segment of the projection, and thus, for each segment distortion is minimized. Nothing has been removed from the map; the outline can be visualized as an orange peel. Any elliptical projection can be interrupted, and Goode's homolosine is almost always in interrupted form, as seen in Figure 5.22. Sometimes interrupted projections are also *condensed*, in which case portions of the earth are removed to allow a larger scale map to be placed on a page. When the interest is in world distributions, population for example, the oceans are of little interest and can be omitted. Normally, the mapmaker indicates that this has been done and where (Figure 5.24).

PROJECTIONS AND MAP READING

When reading a map, you want to know where it is most accurate and what properties it preserves. Is it equal area or conformal? Can distances be measured? Where? Where is the map most accurate? At the equator? The midlatitudes?

For maps of small areas, such as topographic quadrangles, cities and towns, and even small states, the map projection isn't of major concern because there is little recognizable distortion. For larger areas, such as countries, regions, continents, and certainly the world, knowing the projection used and its properties will help you avoid making errors. On world maps and maps of continents and countries, there should be

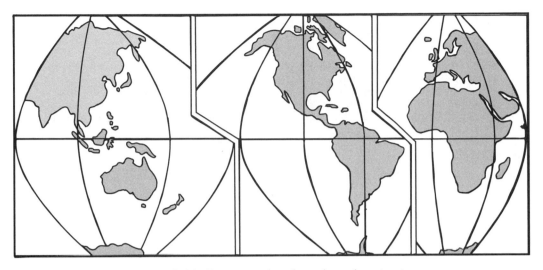

FIGURE 5.24. Interrupted and condensed projection.

(but isn't always) an indication of the projection used. It may be in very small type and hard to find, but if it is there, you can consult a source like Appendix E or Table 5.1 to determine the answers to the questions. But what if there is no indication? Then some detective work is in order. Looking at the parallels and meridians, as well as the size and shape of countries, and comparing to maps on known projections will help.

It is especially important to recognize that map projections are often misused either accidentally or deliberately.

Misuse and Abuse of Projections

Map projections are often misused, as we saw with the Mercator projection. The Mercator is a valuable projection and perfectly suited to its designed purpose of navigation. Unfortunately, it has gotten a bad name because it was used inappropriately in schoolbooks. On propaganda maps, projection misuse is deliberate and is usually intended to mislead the reader, but projections can also be misused through ignorance, inexperience, or negligence. The map reader must be aware and beware of misuse.

One common example of misuse is a nonequal-area map to illustrate a distribution, such as population or crops. Because distributions are area dependent, showing them on a projection that distorts size gives an incorrect impression. Figure 5.25 shows a dot distribution on a Mercator and a Mollweide projection. While the number of dots is the same for each map and the size of the area is roughly the same, the dots appear more sparse on the Mercator projection. The same holds true for comparing the sizes of countries as in the Greenland and Australia examples. Trying to compare distances between cities on a sinusoidal projection is impossible. Distance is measured along a great circle, and this projection doesn't show great circles. The scale on the projection allows measurement only along the equator. While the parallels are drawn to their correct length, the shortest distance between two places on the

TABLE 5.1. Projections "Cheat Sheet"

Equal-area maps

Show areas in correct proportion but distort shapes. Achievable for the entire map. On these, if there is north–south stretching, you must compress the same amount east–west. Used for distribution maps so that areas can be compared. Probably the most used for thematic world maps.

Conformal maps

Show shapes of very small areas correctly. Technically, they show angles with infinitely small sides correctly. Areas are distorted. At the point of intersection, parallels and meridians cross at right angles. On these maps, if you stretch north–south, you must stretch the same amount east–west. Used in navigation. It is not possible to have an equal-area conformal projection; they are mutually exclusive. Arno Peters claimed his projection was, but obviously he was incorrect.

Equidistant maps

Show distances correctly from a single point or along a single line. The azimuthal equidistant is the most common. It is frequently seen in airports. It is centered on a place (LAX, for example), and with the scale you can determine the distance from there to any other point. You CANNOT measure between any other points, however. In this form, oblique case, the grid looks like a plate of spaghetti.

Azimuthal maps

These are all plane projections, and they show azimuth correctly from the center. Azimuth is the angle formed between the observer and the object observed and a true north–south line (meridian) passing through the observer. It is a way of expressing direction, but it has nothing to do with the cardinal directions (NSEW). I have heard faculty describe Mercator as azimuthal; it is NOT. These projections are used in navigation.

Compromise maps

An example is the Robinson projection. Has no specific property—not equal area, not conformal, but looks good and not too far off. Used in atlases and textbooks.

Cylindrical maps

In theory, formed by wrapping a cylinder around a globe and "projecting" the grid. They are rectangular in outline, and have straight meridians and parallels that cross at right angles. There are equal-area and conformal cylindricals. They all have great distortion in the high latitudes because the meridians are parallel rather than converging at the pole as they do on the earth. The zone of best representation is at the equator. In 1999, the various cartographic organizations got together to recommend against using cylindricals for most purposes—especially education. However, for small areas, especially equatorial, they are good.

Conic maps

A cone is wrapped around the globe. The *line* where the cone touches the globe is the standard parallel and the center of the zone of best representation. There are equal-area and conformal conics. They are best for midlatitude areas with east–west extent. They are often used in the *secant* case in which the cone cuts through the globe; in at one parallel and out at another, giving a wider zone of good representation. The United States is most often shown on a conic.

Plane (azimuthal) maps

All of the plane projections are also azimuthal. A plane is tangent to a point on the globe, and the zone of best representation is at the point of tangency. The five plane projections in most common use are the orthographic (old maps of the moon were often on this); stereographic (conformal), old maps showing the world in two hemispheres were usually stereographic; gnomonic, any straight line on the gnomonic is a great circle or great circle arc; equidistant (see above); and equal-area (see above). On all of the plane projections, any straight line through the center is a great circle; on the gnomonic, any straight line is a great circle.

Aspects maps

Any of the developable surfaces can be placed at other parts of the globe. *Cylindricals* are equatorial (wrapped around the equator), transverse (wrapped around a pair of meridians and the poles), or oblique (wrapped around another great circle). *Conics* are normal (touching a parallel); oblique (touching some other small circle); or equatorial (centered over the equator). *Planes* are polar, oblique, or equatorial.

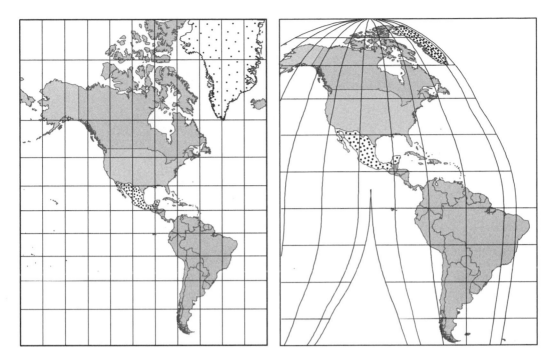

FIGURE 5.25. Dot distribution on Mercator (conformal) and Goode's homolosine (equal area).

same parallel is not along that line, but along the great circle that joins them. Often maps are published that have no indication of the graticule (lines, tick marks); without this guideline, the reader has no way of knowing what can or cannot be shown on the map, or where and what kinds of distortions occur. Again, comparing continent shapes to known projections can help identify the graticule.

A recent type of abuse is the use of the so-called geographical projection that is created by converting longitude and latitude to *x, y* coordinates and is similar to 16th-century plane charts. It is much used on GIS-created maps, but, like other rectangular projections, is highly distorted and not suitable for anything that involves areas or shapes (Figure 5.26).

FURTHER READING

Bugayevskiy, Lev M., and Snyder, John P. (1995). *Map Projections: A Reference Manual.* London: Taylor & Francis.

Deetz, Charles H., and Oscar S. Adams. (1945). *Elements of Map Projection: With Applications to Map and Chart Construction* (Special Publication No. 68; 5th ed. rev.). Washington, DC: U.S. Government Printing Office. (Reprint edition available from Nabu Press, 2011).

Greenhood, David. (1964). *Mapping.* Chicago: University of Chicago Press.

Kennedy, Melita, and Steve Kopp. (2000). *Understanding Map Projections.* Redlands, CA: ESRI Press.

FIGURE 5.26. "Geographical projection." This projection has no special properties and is very distorted.

Klinghoffer, Arthur Jay. (2006). *The Power of Projections: How Maps Reflect Global Politics and History.* Westport, CT: Praeger.

Monmonier, Mark. (2004). *Rhumb Lines and Map Wars: A Social History of the Mercator Projection.* Chicago: University of Chicago Press.

Snyder, John P. (1993). *Flattening the Earth: Two Thousand Years of Map Projections.* Chicago: University of Chicago Press.

Snyder, John P., and Philip M. Voxland. (1989). *An Album of Map Projections* (U.S. Geological Survey Professional Paper 1453). Washington, DC: U.S. Government Printing Office.

RESOURCES

Map Projection Poster (USGS)
   egsc.usgs.gov/isb/pubs/MapProjections/projections.html
Map Projections: From Spherical Earth to Flat Map
   www.nationalatlas.gov/articles

CHAPTER 6

The Earth from Above:
Remote Sensing and Image Interpretation

> Aground, our view is limited to the sides of
> buildings and mountains; but aloft, we behold
> the entire ground plan of artificial and natural
> structures.
>
> —*Mapping* (p. 104)

In Greenhood's day (1944), flying wasn't something most people had done; now, of course, it is a common occurrence. We are accustomed to seeing the earth from above (if the sky is clear and we can peek out of the window during the movie), so seeing roads and houses spread out below us is not an extraordinary sight. Google Earth, Map Quest, and other online mapping services show satellite views of the area we select. The TV weather news routinely shows a satellite view of our area, and newscasters frequently pinpoint the location of a news story on an aerial image. Of course, this wasn't always the case. Until the mid-19th century the only way to get a high view was from a mountain top. Although these air and satellite images are now mundane, do we really understand them or get the greatest benefit from them? In addition to these everyday views, specialized imagery is available and used in many fields, including agriculture, archaeology, astronomy, forestry, geology, and geography and geographic information science. It is also used in real estate development and highway planning. This chapter teaches the basics of *remote sensing* and *visual image interpretation*. We focus on the types of imagery most readers encounter; we are not concerned with the highly sophisticated and technical aspects of remote sensing that require entire books of their own. Although, technically, imagery doesn't fit the definition of maps, many representations, such as online maps and the new US Topo, combine imagery with graphic elements. Thus, the map user needs to know some basics of image interpretation.

Remote sensing is detecting and/or recording information (data) about an object without being in contact with the object. It includes aerial photographs and satellite *imagery*; the term *image* is used because not all of the "pictures" we see are traditional film photographs. *Visual image interpretation* is reading and interpreting basic imagery, whether photographs from an airplane or digital images from a satellite. A *remote sensor* is a device that allows us to gather the information. There are numerous such sensors. The most familiar is the camera, but some others are radar, sonar, X-ray, and thermal infrared scanners. Watching television is a form of remote sensing in that you are seeing images that are beamed into your living room. Remote sensing is done from a *platform*. The platform is the device that carries a camera or other sensor; as we shall see, many devices from kites to satellites to drone aircraft have been used as platforms.

A BRIEF HISTORY OF REMOTE SENSING

The history of remote sensing can be divided into four periods beginning in 1839.

The First Period

The first period is characterized by refinements of sensors and platforms. Experiments in photography go back to at least 1727 when Johann Heinrich Schulze showed that he could make images on a glass plate coated with a mixture of white chalk and silver nitrate; the images were not permanent and had little practical value. It wasn't until the 1830s that Joseph Niepce and Louis Daguerre, working independently, found a way of "fixing" an image so that it would not fade. Daguerre was given credit for the daguerreotype in 1839. The problem with daguerreotypes was that they couldn't be reproduced, but in 1841, William Henry Fox Talbot discovered a way to make a positive copy from a negative.

Early on it was realized that photography could be useful in mapping, and the French began to experiment with it. In 1858, Gaspard-Félix Tournachon, known as "Nadar," took a camera up in a tethered balloon to take "bird's-eye" photographs; he succeeded in photographing a village. The same year Colonel Aimé Laussedat experimented with a glass-plate camera that was carried by a balloon over Paris.

The problem with these early experiments was that the cameras were very bulky and the glass plates had to be developed immediately. This meant that the heavy camera, glass plates, and the complete development equipment *all* had to be taken in the air. A major breakthrough in 1871 was an emulsion that allowed a delay between photography and developing; the plates could be developed after landing. Cameras varied widely in size in the early days, from 2.5 ounces to one that weighed 1,000 pounds and took photographs as large as 4.5 feet × 7.9 feet (1.37 m × 2.41 m).

Platforms also varied. We have mentioned captive balloons that carried the photographer, the camera, and the "dark room," but kites and unmanned balloons were also used. Kites had been used for gathering weather data, but an English meteorologist, E. D. Archibald, is credited with the first successful kite photograph in about 1882. Seven years later a Russian, R. Thiele, connected seven unmanned kites, mounted cameras on them, and sent them aloft for photography. The "panoramagraph" photo could be used for recording cartographic information from remote areas. G. R. Lawrence, an American photographer, who developed the 1,000-pound (453.6-kg) camera, used balloon-kites at heights of over 3,000 feet (914 m). Lawrence was in San Francisco on April 18, 1906, the day of the great San Francisco earthquake and fire; he sent one of his large cameras 2,000 feet (609 m) over the city and produced a large panoramic negative of the city (Figure 6.1). Probably the most unusual platform was first used by Julius Neubronner, who in 1903 designed and patented a breast-mounted aerial camera for carrier pigeons. This camera, the mini-metrogon, made automatic exposures every 30 seconds. The camera was actually used at an international photographic convention in Dresden, Germany, in 1909. Pigeons with cameras harnessed to their bodies flew over the exhibition hall, and the pictures were made into souvenir postcards for the attendees (Figure 6.2).

Although these platforms were creative, they all had one major flaw: They were not easily navigable—certainly a carrier pigeon could not be steered. Although many of the photographs were of excellent quality, they couldn't be made from planned positions and the platforms weren't stable. The invention of dirigibles (blimps) did provide such a platform, but during the first half of the 20th century, the airplane was the most important platform. The first known photographs from an airplane were taken by Wilbur Wright in 1909 over Centocelli, Italy; these were motion pictures. By 1913, air photography for mapping was beginning to be used.

The Second Period

The second period, World War I to World War II, was characterized by increased coverage and the basics of interpretation. Balloons had been used for military observation during the American Civil War, but no photographs are known to have been made then. The use of aerial photographs for military purposes dates to the early years of World War I, and special aerial cameras were being made by the end of 1915.

In the years following World War I, many who had been involved in military photo interpretation became involved with scientific and commercial interpretation. Several journals began to specialize in photogrammetry (making maps from photographs) and photo interpretation. *Photogrammetric Engineering and Remote Sensing*, which is the official journal of the American Society of Photogrammetry (now the American Society for Photogrammetry and Remote Sensing, ASPRS), was first published in 1934 as *Photogrammetric Engineering*.

In the United States, several aerial survey companies were founded in the 1920s and 1930s (at UCLA the Benjamin and Gladys Thomas Photo Archives contains thousands of photos from two of these early companies). Since the 1930s, many U.S. government agencies have made extensive use of aerial photography. Among these agencies were the Agricultural Adjustment Administration which systematically photographed farm and ranch lands over the entire country; the Forest Service, which

FIGURE 6.1. San Francisco in ruins (photo by George R. Lawrence). *Source.* Library of Congress (*www.loc.gov/pictures/resource/ppmsca.07824/*).

FIGURE 6.2. Pigeon with camera. Photograph: wikicommons.

photographed timber reserves; and the U.S. Geological Survey, which photographed areas to make topographic and geologic maps. In the 1930s, the Tennessee Valley Authority and other regional agencies began to use photography for planning purposes, as did state, county, and metropolitan planning agencies.

World War II was a major stimulus to photo interpretation. The Germans, Japanese, and British all had programs of photo reconnaissance and photo intelligence; the Americans developed theirs rather late. The United States had almost no capability in military photo interpretation when it entered the war in 1941, but, soon thereafter, in January of 1942, the Navy had founded a Navy Photographic Interpretation School.

The Third Period

The third period, 1945–1960, was characterized by refinement of analytic techniques. When veterans returned to civilian life as geologists, engineers, foresters, geographers, soil scientists, and archaeologists, they found many civilian applications of photo interpretation. Some developed courses at universities, and remote sensing as an academic discipline dates to the early 1950s (at this time the field was called air photo interpretation). During this period, interpretation advanced from simple recognition of features to analysis of photos, changes through time, and regional patterns. Textbooks on photo interpretation were developed, and the period was capped by publication of *The Manual of Photo Interpretation* in 1960. The developments in the later 1960s were rudimentary or even unheard of at the time of its publication.

The Fourth Period

The fourth period, 1960 to the present, marks the birth and rise of the broader field of remote sensing. The term *remote sensing* was coined that year by Evelyn Pruitt who was employed by the Office of Naval Research, and it soon gained wide acceptance. That year also marked the beginning of systematic satellite observations of the earth. *Tiros I*, the first meteorological satellite, was launched that year and provided a series of color images taken by an automatic camera on the unmanned spacecraft. *Gemini III*, the first manned Gemini flight, was a major impetus. Although the astronauts did not carry out a formal photographic experiment, they had a hand-held Hasselblad camera; the photographs taken by that mission generated much excitement from both scientists and the public, and truly marked the beginnings of earth orbital photography. Photographs from weather satellites that we now take for granted were greeted with oohs and aahs. Photographs of the Nile Delta and California's Salton Sea were dramatic (Plate 6.1), and when an image of the earth taken from the moon was first available, it appeared on the cover of virtually every new earth sciences textbook (Plate 6.2). The possibilities of this new technology could be seen, and it is hard to overestimate the excitement and interest. It might be called the "gee whiz" period.

Radio-controlled drones equipped with cameras represent a quite recent platform for remote sensing. Such "unmanned aerial vehicles" (UAVs), or drones, range in size and can fly at very low altitudes, making them useful for both surveillance and mapping.

By the middle to late 1970s, people had begun to take satellite imagery for granted, and many practical applications had been found. The Landsat satellites were specifically designed to provide imagery of the earth on a regular basis. Now, such imagery is routine, and satellite views are available from many sources, including Google Earth, on our home computers, tablets, and even smart phones. Accordingly, being able to understand the rudiments of remote sensing and visual image interpretation should be a basic skill for those in geography and those who use maps on a regular basis.

USING REMOTE SENSING

Four important terms in this chapter are photogrammetry, visual image interpretation, imagery, and remote sensing. *Photogrammetry* is the science of making measurements from imagery and making maps using the imagery. This chapter covers the basics of *remote sensing* and *visual image interpretation*. Remote sensing, which we have already discussed, may be defined as "the art and science of obtaining information about an object without being in direct physical contact with it." It includes aerial photographs and satellite *imagery*. Here we will focus on visual image interpretation because it requires little in the way of equipment and is the type of interpretation that most people will do. This kind of interpretation can be done with conventional photographs or satellite imagery, including the images on sites such as Google Maps and Google Earth. Because remote sensing is a separate discipline and requires specialized equipment, it cannot be covered in this brief space; several excellent texts and resources are listed at the end of this chapter. The American Society of Photogrammetry and Remote Sensing is the major professional organization for this field.

We need to distinguish between *active* and *passive* sensor systems. A passive system simply records the energy emitted by or reflected as radiation by the object itself. The sensor does not transmit any energy to the object. The most common example is a camera, which simply records the light reflected from the object. By contrast, active systems are those that emit their own electromagnetic energy and then record the energy that is reflected back to the sensor. If we take a photograph in a dark room using flash, we have used a simple active system. Radar is a more sophisticated example of an active system. In the early days of the field, the camera was the only type of sensor and a photograph was the resulting image. Now we use the terms *sensor* and *imagery* since there is such a variety of devices and not all produce "photographs."

Sensors "sense" in different portions of the electromagnetic spectrum. *Electromagnetic energy* is all radiant energy that moves with the constant velocity of light in a harmonic wave pattern. The *electromagnetic spectrum* is an ordered array of known electromagnetic radiations extending from the shortest cosmic rays through gamma rays, X-rays, ultraviolet radiation, visible radiation, infrared radiation, microwave, and all other wavelengths of radio energy (Figure 6.3).

The electromagnetic spectrum is usually thought of as being divided into regions or bands distinguished from one another by wavelength or frequency. *Wavelength* is the distance from crest to crest of waves; *frequency* is the number of waves to pass a given spot in a specified time (Figure 6.4). Since all electromagnetic energy travels at the speed of light (186,000 miles per second [300,000 km per second]), longer wavelength energy will have fewer crests passing a point per unit of time than shorter wavelength energy. Therefore, the greater the wavelength, the lower the frequency. Wavelengths vary from hundreds of kilometers down to billionths of millimeters. Thus, the electromagnetic spectrum is actually a continuum, and the division into bands or regions is one of convenience since various sensors only "see" certain portions of the spectrum.

The *visible spectrum* is the portion of energy that the human eye can see. It is only one small portion of the entire electromagnetic spectrum. The visible spectrum includes energy with wavelengths from about 0.4 micrometer to nearly 0.8 micrometer. (1 micrometer is 1 millionth of a millimeter, also called a micron, symbol μm.) What we call colors are the different "bands" of the visible spectrum; a rainbow is the visible spectrum arrayed in wavelength order, red, orange, yellow, green, blue, and violet. A useful mnemonic is ROY G. BV (this used to be ROY G. BIV, but indigo was removed; Figure 6.5).

The conventional camera as well as the eye senses in the visible region. Next to the visible portion we find *near infrared* on the longer wavelength side and *ultraviolet* on the shorter wavelength side. Although we cannot see infrared with our eyes, both film and digital cameras can sense the near infrared with the use of special film or special filters. Cameras can also sense in the ultraviolet portion of the spectrum, again with special filters.

WAVELENGTH

GAMMA RAYS	X-RAYS	ULTRA-VIOLET	VISIBLE	NEAR INFRARED	SHORT INFRARED	MIDDLE INFRARED	THERMAL (RED) INFRARED	MICROWAVE	RADIOWAVES VHF TO LF

ELECTROMAGNETIC SPECTRUM

FIGURE 6.3. The electromagnetic spectrum.

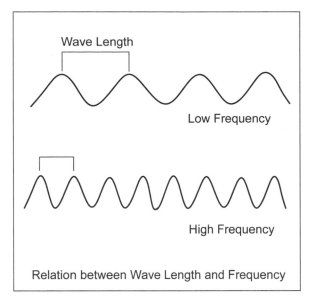

FIGURE 6.4. Wavelength and frequency.

VISUAL INTERPRETATION

Figure 6.6 is an air photo of the harbor at Long Beach, California, and Figure 6.7 is a topographic map of the same area. The photograph was taken from an aircraft with the camera pointed straight down; it is called a *vertical* photograph. If the camera is tilted, the photo is said to be *oblique*—either *high oblique* if the horizon is visible or *low oblique* if the horizon is not visible. Most visual interpretation is made from vertical imagery. The topographic map was made with the assistance of imagery, probably air photos, given the publication date of the map.

Basic interpretation of imagery is straightforward, and several steps are involved. The first step is to determine the scale of the photo. Unless a photo has been *rectified*—that is, adjusted so that the scale is uniform across the image—the scale will vary across the photo depending on how close the area is to the camera and the camera focal length (Figure 6.8). Thus, the scale on unrectified photos is called the *nominal scale*. This scale can be found in several ways. In some cases the scale is provided with the photographs, but many times the user must determine the scale. For con-

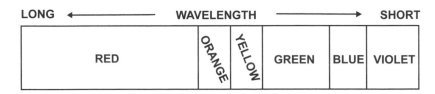

FIGURE 6.5. The visible spectrum.

FIGURE 6.6. Harbor at Long Beach, California. *Source.* USGS.

ventional air photos, the scale can also be determined by the ratio between the focal
length of the camera and the altitude of the plane if these measurements are known.

A common way to find a photo's scale is from a known distance; if the distance
between two points is known, such as a road between two crossroads, one measures
the photo distance, and by using the techniques from map scale, one can create a
scale by relating photo distance to actual distance:

$$\text{Photo distance/distance} = 1/x$$

A third way of determining scale is similar, but uses the size of known standard
features on the photo. Baseball diamonds, tennis courts, soccer and football fields
have standard measurements. The distance between bases on a baseball diamond is
90 feet. Thus using the formula above, measure the distance between bases on the
photo and compare it to 90 feet to determine the scale. Some standard dimensions
are shown in Table 6.1.

After determining the scale of the imagery, we want to identify features. A number of clues can help us to do this: size, shape, shadow, color and tone, texture and
pattern, site and situation, and associated features and objects. These clues are used
together to interpret an image (see Table 6.2).

Size of features is quite simple to determine once one knows the scale of the

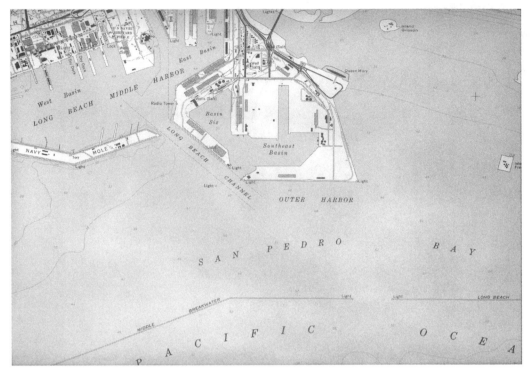

FIGURE 6.7. Topographic map of same area of Long Beach, California. *Source.* USGS.

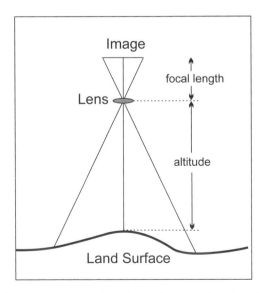

FIGURE 6.8. The scale of a photograph is dependent on the elevation of the platform and the focal length of the camera.

TABLE 6.1. Image Interpretation: Clues to Object Recognition

Clue	Information
Size	How large or small a feature is, as determined from the photo.
Shape	Not only basic shapes such as squares and circles, but irregular shape. Churches are often in cruciform shape. The Pentagon is an easily recognized building from its shape.
Shadow	Tall features can often be identified by the shadows they cast, and their height can be determined from the shadow.
Tone/color	Many features have distinctive colors, such as swimming pools, but even on black-and-white images, shades of gray can identify features.
Pattern	Pattern is the arrangement of features. Orchards normally have a regular pattern; naturally occurring vegetation is random. Contour-farmed row crops tend to have a series of parallel curves.
Texture	Texture refers to the density of a pattern. Orchards have a coarse pattern, row crops a smooth pattern.
Location	Is the feature on flat land or a slope? Is it near transportation or water features?
Associated features	Farms have barns, silos, and fields; types of industries can often be identified by associated features such as storage tanks, rail yards, and docks.

photo, but one can also determine relative sizes to distinguish houses from large buildings, or shrubs from trees, or a thruway from a country road.

With a little experience, features can be recognized from their *shapes*, although you must remember that on a vertical image all features are shown in plan view and it can take some practice to visualize objects in this way. Most of us would not recognize our own home in this view. For practice, you can find a familiar place on the satellite feature on Google Maps and zoom in to identify features. Using your own neighborhood, you can see the shape of your house from above as well as other features, such as schools and shopping centers.

Many objects that are difficult to identify in plan view can be recognized quite easily from their *shadows*. Examples are water towers, microwave towers, monuments, storage tanks, and some trees. We can determine the height of such tall features from their shadows if the time of day is known (Figure 6.9). Shadows can even help determine the number of floors in a building.

Color, *tone*, and *texture* are quite useful. Color photographs are somewhat easier to interpret than black and white, especially for the novice, because many objects can

TABLE 6.2. Features with Standard Dimensions

Baseball diamond	Home plate to pitcher's box: 60 feet 6 inches Distance between bases: 90 feet
Tennis court	120 feet by 60 feet
Football field	100 yards plus 10 yards for each end zone
Soccer field	Minimum 100 yards by 50 yards Maximum 130 yards by 100 yards International 110 yards by 70 yards

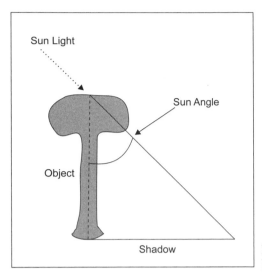

FIGURE 6.9. The height of a tall feature can be determined from the length of its shadow.

be recognized from their color alone, such as backyard swimming pools. On black-and-white images, some practice is needed to interpret the tones (shades of gray), but the interpreter quickly learns that a dirt road looks different from a paved road and that a wheat field ready for harvest is lighter than one that is not. *Texture* refers to the apparent roughness or smoothness of features. Thus, on a golf course, the greens and fairways have smooth texture, but the rough, appropriately, has a rough texture. Scrublands have a bumpy or grainy look, and grasslands appear smooth. It should be kept in mind that these three clues—color, tone, and texture—can vary with time of day or weather conditions. Water can appear light or dark depending on the sun angle and rough or smooth depending on the wind. During World War II, infrared film (called camouflage detection film at that time) was invented. Because this film showed live vegetation differently from dead vegetation, bunkers, trenches, and other military features could be identified because they were often covered in cut tree limbs. After the war, environmentalists made extensive use of color infrared film to identify dying plants and distressed vegetation; healthy vegetation appears red on an infrared photo (Plate 6.3).

Pattern, or arrangement of features, is important in both map and image interpretation. Many features have distinctive patterns, such as roads, rivers, and fields. Orchards have a different pattern and texture than row crops such as corn fields. Residential patterns can help distinguish rural areas from urban. Patterns of lines can identify parking lots, football fields, and tennis courts. Patterns can tell you where you are when you look out of an airplane window: The cadastral patterns of French Long Lots tell you that you are looking at a French-settled area; rectangular survey patterns say Midwest; and the sharp demarcations at the Canadian and Mexican borders where survey systems change between countries show a boundary "line," even though it is not drawn on the land or the photo. Physical patterns of terrain, vegetation, and drainage show the Appalachians and the Rocky Mountains. Looking at satellite views of areas on Google is a good way to see patterns and learn how to recognize them. For example, along the Mississippi River near New Orleans, we can see long lots, and in Walcott, Iowa, the PLSS pattern is obvious. Road patterns, population density patterns, and types of vegetation can all be seen by "traveling" around the world.

Location as used here is more than latitude and longitude; it also includes site and situation. Is the feature on a hill, a slope, or a valley? Is it in an industrial area, a residential area, or a rural area? These can help identify not only individual features, but also groups and types of features.

Farms are normally accompanied by outbuildings—barns, silos, and animal pens; oil refineries have pipes, tanks, and towers; and shopping malls have large parking lots. These *associated* features can help interpret the characteristics of an area.

With the introduction of the new topographic map type, US Topo, which combines imagery and map views, we find that image interpretation is becoming more important for the map reader, as we shall see in Chapter 9.

FURTHER READING

American Society of Photogrammetry. (1960). *Manual of Photographic Interpretation*. Washington, DC: Author.

Campbell, James B., and Randolph H. Wynne. (2011). *Introduction to Remote Sensing* (5th ed.). New York: Guilford Press.

Jensen, John R. (2006). *Remote Sensing of the Environment* (2nd ed.). Englewood Cliffs, NJ: Prentice-Hall.

Johnston, Andrew K. (2004). *Earth from Space*. Buffalo, NY: Firefly Books.

Lillesand, Thomas, Ralph W. Kiefer, and Johnathan Chapman. (2007). *Remote Sensing and Image Interpretation* (6th ed.). New York: Wiley.

Paine, David P., and James D. Kiser. (2012). *Aerial Photography and Image Interpretation* (3rd ed.). New York: Wiley.

Reeves, Robert G. (Ed. in Chief). (1975). *Manual of Remote Sensing: Volume I. Theory, Instruments, and Techniques*. Falls Church, VA: American Society of Photogrammetry.

Reeves, Robert G. (Ed. in Chief). (1975). *Manual of Remote Sensing: Volume II. Interpretation and Applications*. Falls Church, VA: American Society of Photogrammetry.

RESOURCES

American Society of Photogrammetry and Remote Sensing (ASPRS)
www.asprs.org
Journal: *Photogrammetric Engineering and Remote Sensing* (PE&RS)
Earth Resources Observation and Science Center (EROS)
www.eros.usgs.gov
Google Earth
earth.google.com
National Aeronautics and Space Administration (NASA)
www.nasa.gov
Natural Resources Canada
www.nrcan.gc.ca

PART II

Map Types and Their Analysis

CHAPTER 7
Virtual Maps and GPS

The use of flight and light, electronics,
and of surpassingly productive mechanical
mathematicians in the making of today's maps
is putting into the hands of both the advanced
scientist and the first-grade child cartography
better than any dreamed of before.

—*Mapping* (p. xiii)

It is doubtful that nearly 50 years ago Greenhood anticipated the magnitude of the changes of the 21st century or the widespread use of digital technology. There were no personal computers, computer tablets, smart phones, or global positioning satellites. And while the maps are not all "better" than any before, certainly they are more available and come in a wider variety. Virtual maps are now so widespread that we forget that maps on the Internet are still quite new, a product of only the past two decades: MapQuest dates to 1996 and Google Maps was first used in 2005. These maps have antecedents in early computer maps that could only be viewed as crude paper printouts from line printers (Figure 7.1, SYMAP), but now we can search for a restaurant, get driving, public transportation, walking, or bicycling directions to it, view the map with traffic conditions, and see a picture of the restaurant so that we know we have arrived at the right place even without leaving our automobile.

MAPS ON THE INTERNET

Maps on the Internet are in two forms: *static* and *dynamic*. Static maps are essentially conventional paper maps that have been scanned and posted online. The his-

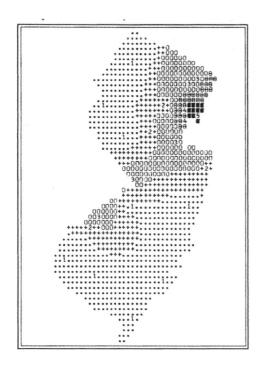

FIGURE 7.1. SYMAP.

toric maps of David Rumsey's website are examples (Figure 7.2). The online maps appear smaller than they actually are, but the sites usually allow the user to zoom in and out and pan across the map. Nothing is changed on the original map, although some sites will allow simple drawing, highlighting, or annotation.

Dynamic maps are *interactive*; that is, the user can click on a point and get information about the point—in other words, query the map. Dynamic maps may also be animated and have sound effects, pictures, or texture; that is, they are *multimedia* or *multisensory*.

The type of interactive webmap most commonly encountered is the navigation map of the type available through Google Maps, MapQuest, Bing, and AAA Trip-Tiks (see also Chapter 10). These are all quite similar and even use the same databases with the exception of Google. One can query these maps to a limited extent; that is, pointing at a place and right clicking with a mouse will give the address and perhaps some additional information about the place. The maps allow a strictly map view, or they may be overlain on a satellite map; some allow a "bird's-eye" or oblique view of the satellite version or a vertical view. Street views of places are increasingly provided with the maps. Zooming in and out with these webmaps changes the graphic scale that is included on the map, allowing the user to estimate or even measure distances (Figure 7.3).

The "satellite views" of such maps are made up of thousands of images, some of which are high-resolution aerial photography taken from altitudes of 1,500 feet and others are actually from satellite images. The images are updated regularly but may still be a few years out of date; the resolution of the imagery (amount of detail) varies depending on the area. Rural areas may be at lower resolutions than urban areas;

urban areas are more likely to be imaged from high-resolution aerial photographs. In addition, some areas considered of strategic importance may be deliberately blurred.

There are also occasional errors on these webmaps, often misplaced addresses, incorrect street names, or even misplaced streets. The user can contact the database provider and correct these errors. At the bottom of the image, the map data provider is listed: GoogleMaps, NavTeq, or Tele Atlas. Their websites have provision for users to submit map corrections. It is a good idea to submit the correction to all of the providers unless you know that only one is inaccurate. Although these maps are provided through the Internet and are usually viewed on a computer screen, the basics of map reading—scale, direction, latitude, and longitude—apply just as for paper maps. Of course, they can also be printed as paper maps. Also, please note that all of these Internet map sites are revised frequently.

GOOGLE EARTH

Google Earth is a virtual globe, map, and GIS program that maps the earth by superimposing satellite imagery, aerial photography, and a GIS three-dimensional globe. It is a commercial product with multiple applications that also allows users to add their own data and create maps. In addition to allowing users to take tours of various places, it permits searches much like Google Maps and provides an image of the place with its coordinates and elevation. It even allows the user to view older imagery, if it is available, to get a historic view. The company has also partnered with the David Rumsey collection of historic maps to combine historic maps with Google Earth.

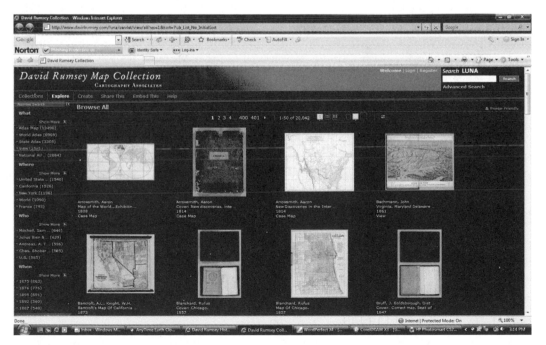

FIGURE 7.2. Screenshot of a browse page on David Rumsey's website. From *www.davidrumsey. com*. Reprinted by permission.

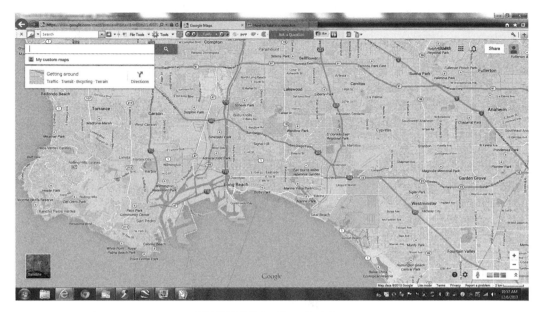

FIGURE 7.3. Google Map.

GOOGLE MOON, GOOGLE MARS, AND GOOGLE SKY

These sites work in much the same way as Google Earth and provide imagery, maps, and historic information about the planets and the stars. For example, Google Moon, which was introduced on July 20, 2009, to celebrate the 40th anniversary of the *Apollo 11* mission when humans first walked on the moon, shows the sites of all *Apollo* landings with information about the mission and the landing site. Elevation maps are available for both Google Moon and Google Mars as well as conventional and infrared imagery (Figure 7.4). Google Sky uses imagery from the Hubble Space Telescope.

ANIMATED MAPS

Animated maps create the illusion of change, either in time or space. They accomplish this illusion by rapidly displaying single maps so fast that the eye cannot detect the individual images. Animated maps have a long history dating to before World War II, when they were used in the newsreels of the time to show movement of troops and the position of the front. Although their potential was recognized, they were not widely used until the introduction of desktop computers because they were time consuming to produce and required special equipment—movie projectors—to display. Since the 1990s, cartographers have explored the possibilities of animated maps, especially online. Animated maps range from simple base maps with moving symbols to full multimedia presentations (see below) with sound and even interactivity.

There are two basic kinds of animated maps: *temporal*, which show change over time, and *nontemporal*, which show change in space of some attribute. Many maps are both. Temporal maps might show the spread of a drought, changing climates over time, or the spread of a disease. Spatial change includes so-called flybys or fly-throughs that give the impression of flying over an area. Google Earth has this capability. Animated temporal maps include, in addition to the basic map elements, time scales and legends that indicate hours, minutes, days, or months. Most now include pictures, sound, and dynamic symbols, and on most, the reader can stop action to examine a particular map more closely. Some simple animations of earthquakes can be seen at *http://earthquake.usgs.gov.*

Many animated atlases on the web show such subjects as American territorial growth, the Holocaust, and the Civil War. Because this book can only display static images, a few animated websites are listed under Resources at the end of this chapter.

SOUND MAPS

Sound is used in a variety of ways on webmaps and other electronic maps. It is used to tell a narrative as on many animated multimedia maps; as a symbol, such as bomb blasts, horns, or whistles; to reinforce a label, such as a spoken place name as well as a written name; to replace labels so that the map is not cluttered; and to sound an alarm to warn the user of an error. Sound symbols may be abstract or realistic. Narration, car horns, and the like are realistic, but beeps and other warning alarms are considered abstract. In some cases, we are not really aware of the use of sound; we are so accustomed to narration on television and motion pictures that the sound is "transparent."

FIGURE 7.4. Google Moon.

HAPTIC MAPS

Haptic maps refer to the use of touch with maps. Touch has been used for maps for the blind: *Tactile maps* have raised symbols that the user can feel with the fingertips, much like braille writing. But the use of touch with electronic maps is a new area. Even though touch-screen monitors and tablets are available, the screen is smooth and cannot be felt in the way a conventional tactile map can. Haptics is associated with the "feel" of objects, such as heat and cold, rough or smooth. The sensations are obtained through a mouse, wand, or even a special glove, and haptics can be used for maps for both sighted users and those with visual impairments. This is a very new area that has great potential.

MULTIMEDIA MAPS

Multimedia maps can also be called multisensory maps because the maps draw on our senses of sight, hearing, and touch. These maps use the computer not only as a tool, but also as the medium of delivery, and they have multiple levels or layers that allow the user to explore more deeply. For example, a simple map of an area may allow the user to roll the mouse over the screen and produce pop-ups of information, including photographs and videos. Or, increasingly, with touch screens and tablet computers, the user touches at various points on the screen to access information.

As with other maps, these maps are only as good as their maker and the data used. It is easy to get drawn in by the "gee whiz" factor, to be impressed with the bells and whistles, the attractive colors, the clever animations, and sound effects, but if the display is presenting erroneous or even deliberately false information these maps, like any map, can be dangerous. The answer, of course, is to be an informed map user and not rely on any single map.

GLOBAL POSITIONING SYSTEM

GPS is a satellite-based navigation system that allows the user to determine location, speed, and time anywhere in the world. The system is made up of three segments—satellite, ground control, and user—and can send and receive radio signals. The service is provided by the U.S. government at no charge, and the Department of Defense maintains the system (Figure 7.5).

GPS burst on the public scene some 25 years ago and has become so widespread that it has even become a metaphor for navigating one's life or organization. Books have been published on GPS for careers, GPS for companies, and GPS for spiritual enlightenment. None of these uses of the term refers to the tool used for spatial navigation.

The idea for GPS is not new; the concept goes back to the earliest satellites in the 1960s. In 1973, the forerunner of GPS was created; Navstar-GPS was originally intended for military use, but President Reagan directed that it be made freely available for civilian use after it was fully developed. The first satellite for this phase was launched in 1989 and the 24th in 1994. Originally, the highest quality signals

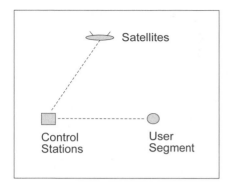

FIGURE 7.5. GPS segments.

were restricted to the military; this was called selective availability. President Clinton ordered selective availability turned off at midnight on May 1, 2000, which improved accuracy for civilian uses.

The first segment of GPS is the *space segment*, which consists of at least 24 satellites that orbit the earth twice a day at about 7,000 miles per hour at a height of 12,000 miles. At any given time, any GPS receiver can receive signals from at least four satellites. The satellite signals can penetrate clouds, glass, and plastic, but solid objects decrease the power of the signals and the signals cannot pass through objects that have a lot of metal or water. Thus, GPS cannot be used in caves or under water. The *control segment* constantly monitors the satellites from ground stations. These stations check the health, signal, and orbit of the satellites.

The *user segment* is the part we are concerned with here. The user segment is the GPS receiver, which collects signals from the satellites that are in view. If three satellites are viewable, the unit can calculate latitude and longitude, and with four or more, the altitude can be determined. From the receiver's position, the GPS can then calculate such information as speed, bearing, trip distance, track, and sunrise and sunset times. Generally, the systems are accurate to within an average of 15 meters, but newer units have WAAS (wide area augmentation system) capability and are accurate to within an average of 3 meters. No information is sent from the user's GPS unit back to the satellite.

The first GPS units, and some specialized units, do not have maps built in. Some are designed to be worn like a wristwatch and to give location and trace a track. Basic bicycle and runner's GPS units provide location, speed, and distance. GPS receivers in automobiles and for outdoor activities usually have preloaded maps (Figure 7.6).

GPS applications are widespread. The most obvious applications are navigation for planes, ships, and automobiles. Farmers use the technology to plow in GPS-guided tractors, map fields, harvest, and apply fertilizer and pesticides. It is valuable for surveying disaster areas and mapping rapidly moving events such as oil spills, forest fires, and hurricanes. GPS units are used by hikers, fishermen, bicyclists, golfers, geocachers, and runners to navigate and to mark and record places and routes. Of course, the military is a major user, with equipment mounted in aircraft, ships, tanks, and other vehicles for navigation, air support, and target designation. Emergency vehicles use GPS to locate accidents. Delivery trucks and vans that bring packages to your home use GPS to find addresses. Surveyors use GPS for everything from

FIGURE 7.6. Automobile GPS unit.

delineating property lines to designing the infrastructure of an urban area. (The use of GPS for automobile navigation will be treated in Chapter 11 and discussed with topographic maps in Chapter 8.) The rise of GPS marked the decline of a major map type—the street map and especially the street atlas. Delivery trucks and vans in urban areas typically had a battered street atlas on the dashboard (in California the *Thomas Guide* was ubiquitous), but with GPS, the driver simply punches the address to the unit and no longer carries his "Tommie Guide."

There are caveats when using GPS. For example, errors can affect the accuracy of the signal. Starting the unit in an area with an unobstructed view of the sky will reduce many of these errors. Among these errors are delays as the signal passes through the atmosphere, although the system has a built-in model that corrects some of this type of error. There may be errors in the satellite's reported location; signals can bounce off of objects such as tall buildings or large rock surfaces, delaying the signal and causing errors. Thus, using a GPS in deep canyons can present problems. The more satellites the receiver can view, the more accurate the position.

The biggest cautions must be attached to the maps. It is important to understand that maps and other information, such as roads and points of interest, are not part of the GPS system; they do not come from the satellites, and the U.S. government's GPS program does not provide any of that data. Map information on GPS receivers is provided by the commercial companies that manufacture the units. The companies get their digital maps and data from the same digital mapmakers or digital content suppliers as do MapQuest and other Internet map providers, primarily NavTeq and TeleAtlas. The GPS coordinates may be correct, but the map might be out of date. Maps require constant updates to stay accurate. Roads change. New roads are built, old roads are rerouted or closed, temporary detours are made. The points of interest (POI) on many systems show restaurants, banks, and stores, but new ones open and old ones close. The map doesn't magically update itself; the user must install new maps, ideally once a year. Some units come with very basic maps, suitable for navigating through cities or along highways; however, more detailed maps can be downloaded. Those GPS units designed for hiking and other outdoor activities may come preloaded with topographic maps with contour lines and shading; they may also allow a bird's-eye or three-quarters view of the landscape (Figure 7.7).

In addition to currency issues, the maps may have other errors. A dirt road may look like a paved road, or an address may be shown in the wrong location. The consequences of these errors can be as minor as having a package delivered to the wrong address, or as major as the map user getting mired on a dirt road or being arrested for trespassing on a private road. Frequently, newspaper articles describe a driver or hiker getting lost while depending on the GPS; these experiences range from humorous to tragic. (Some of these experiences will be discussed in Chapters 8 and 10.) The user should not put blind faith in the tool, should be familiar with the features and uses of their particular unit, and should use common sense. If you find an error on your GPS map, as with Internet maps, you can report it to the map suppliers, Google, NavTeq, and Tele Atlas.

GPS receivers, even those without built-in maps, allow the user to locate points called waypoints and mark them in the receiver's memory; the receivers will trace routes on the unit by joining the waypoints. This allows the user to retrace the route to the origin or to plot it on a paper map or computer. The unit can be used as a compass to provide directions. More sophisticated units will draw profiles of the route, and if there is a built-in map, it will trace the route on the map.

Paper maps and GPS complement one another. Paper maps provide an overview of an area and show landmarks, roads, waterways, and landmarks; GPS lets you identify a point anywhere in the world, navigate to it, and create a path to the destination. Garmin, Inc.'s downloadable manual on using GPS with paper maps for land navigation provides an excellent introduction to combining maps and GPS.

FIGURE 7.7. GPS for hiking.

Because the technology of all of the maps and tools on this site is changing rapidly, readers are strongly urged to consult the current manuals and help features available for webmaps and GPS. Websites for some of these resources are listed below.

FURTHER READING

Cartwright, William, Michael P. Peterson, and Georg Gartner. (2007). *Multimedia Cartography* (2nd ed.). New York: Springer.

Easton, Richard D., and Eric F. Frazier. (2013). *GPS Declassified: From Smart Bombs to Smartphones*. Lincoln, NE: Potomac Books.

Peterson, Michael P. (1995). *Interactive and Animated Cartography*. Englewood Cliffs, NJ: Prentice-Hall.

RESOURCES

Animated Atlas
www.animatedatlas.com
Civil War Maps
www.civilwaranimated.com
Historical Maps
www.the-map-as-history.com
United States GPS site
www.gps.gov
Garmin GPS
www.garmin.com
NAVTEQ Maps
www.navteq.com
Google Maps
www.maps.google.com
Garmin, *An Introduction to Using a Garmin GPS with Paper Maps for Land Navigation*
www.garmin.com/manuals/usingagarmingpswithpapermapsforlandnavigation

CHAPTER 8

Topographic Maps

What you'd have would be the best kind of relief
map so far invented [contour lines] to give on flat
paper all the facts about the wrinkled, dimpled,
pimpled, fissured, dented, swollen, hollowed, roly-
poly surface of the earth.

—*Mapping* (p. 78)

As described by the U.S. Geological Survey (USGS), "Topographic maps usually por
tray both natural and manmade features. They show and name works of nature
including mountains, valleys, plains, lakes, rivers, and vegetation. They also identify the principal works of man, such as roads, boundaries, transmission lines, and
major buildings" (USGS booklet *Topographic Mapping*). What most distinguishes
topographic maps are the contour lines used to symbolize elevation and show both
the elevation and shape of the land. As noted in Chapter 3, topographic maps are
one of the most widely used kinds of maps. They are used by engineers and planners
for public works design, for residential and commercial development, and for energy
exploration. They are valuable in environmental management and conservation, and
they are used by the general public in outdoor activities such as hiking, camping, fish-
ing, orienteering, and geocaching.

Although many countries produce topographic maps, there are great similarities in all topographic maps. With the aid of a symbol sheet, a map reader who is
familiar with the maps of one country can read the maps of almost any other country with some confidence. For that reason, in this chapter we focus on maps of the
United States. While an abbreviated symbol sheet is reproduced here (Plate 8.1), it
is strongly recommended that the reader download the *Topographic Map Symbols*

booklets from *http://egsc.usgs.gov/isb/pubs/booklets/symbols/topomapsymbols.pdf* and *http://nationalmap.gov/ustopo/images/US_Topo_Symbol_Sheet* as a reference while reading this chapter.

A BRIEF HISTORY OF USGS TOPOGRAPHIC MAPS

In 1884, 5 years after the USGS's founding, John Wesley Powell, the director of the USGS, asked Congress for authorization to begin systematic topographic mapping of the United States. In the earliest days, the maps were largely created in the field by surveyors who would sketch the contours by eye after measured control points had been developed. The earliest maps were at a scale of 1:250,000 for 1° areas and 1:125,000 for 30' areas. By 1894, most maps were at a scale of 1:62,500 and covered 15' quadrangles. The current most common scale is 1:24,000 and 7.5' quadrangles. Officially, 1992 marked the completion of coverage of the entire conterminous (48 states) United States, and these maps are no longer being made.

Aerial photography and photogrammetry, following a period of experimentation in the 1920s, was implemented for mapping the Tennessee Valley Authority's projects in the 1930s. This allowed more accurate drawing of contour lines by viewing a three-dimensional view of the land through stereo pairs of photographs.

As with many mapping activities, computers have had a major impact on topographic mapping. They were first used in the 1960s to eliminate tedious hand drawing, but most of the work is now done in the lab. As we will see, the newest generation of topographic maps are digital and viewable from the personal computer.

In 2009, the USGS introduced a new type of topographic map that takes advantage of digital technology. Called *US Topo*, these maps are in the 7.5' quadrangle format and can be downloaded free, or paper versions can be purchased from the USGS Store. Coverage of the 48 conterminous states, the "lower 48," was completed at the end of 2012; Hawaii and Puerto Rico were completed in 2013, and Alaska is nearly complete. Downloaded versions can be printed from personal computers or plotters as full-sized maps, or maps can be customized in a format specified by the user, such as a map segment. The maps are made up of layers of geographic data: orthoimagery (rectified images), geographic names, elevation contours, hydrographic features, boundaries, transportation features—all of the information shown on a conventional topographic map overlaying an orthoimage. These layers can be turned on and off when viewing on a computer or printing the map. This feature is especially useful when analyzing and interpreting the maps. The user can turn on only roads or only contour lines, for example. A User's Guide is available that includes a quickstart page (see Resources at the end of the chapter). While this system of delivering topographic maps has many advantages, including cost effectiveness, the program has been criticized because the maps are not field checked for accuracy and features not included in public domain databases will not appear on the maps. Thus, features such as mines and oil wells will not be named on the maps. The trade-off is that these maps can be updated more quickly. Conventional maps are still available through USGS as historic topographic maps.

The basics of reading topographic maps are much the same for both conventional and US Topo maps, but where there are differences, they will be noted.

USING TOPOGRAPHIC MAPS

Marginal (Collar) Information

The space outside the neat line of the map identifies the map and gives other useful information (Figure 8.1). There is some conflict in terms on the new US Topo series. On earlier maps this area was called the margin, but on these new digital maps it is called *the collar*. All of the information except the orthophoto within the neat line on these maps is referred to as the frame information. These differences must be kept in mind when viewing US Topo online. The first piece of information is the title, which is the quadrangle name, and is found in the upper right margin. Other information that might be shown here include the state name and the series type. In the lower right margin is another title block that again gives the name, state, and date, but also gives the geographic index number, which is the coordinate of the corner of the map nearest the Greenwich meridian and the equator.

If the map has been made by another agency in cooperation with USGS, that information is placed in the middle of the top margin. In the upper left margin is the name of the publisher, which is always

UNITED STATES
DEPARTMENT OF THE INTERIOR
GEOLOGIC SURVEY

Adjoining quadrangle names are shown in the margins where the respective quads would touch. Geographic coordinates are shown at all four corners and at 2.5' intervals on the 7.5' quads. The coordinates for the state and UTM grid systems are shown near the corners.

In the lower left margin is a *credit legend* that lists the name of the mapping agency, the name of the editing and publishing agency, the name of the agency or agencies that furnished the geodetic control, the method(s) by which the mapping was performed, a credit note for hydrographic information, and any explanatory and informative notes.

Taken together, the marginal information performs the function of the title page, copyright page, and table of contents for a book.

Series and Scales

Topographic maps are large-scale maps that cover small areas in great detail. In the United States, the basic scale for these maps is 1:24,000, or 1 inch represents 2,000 feet. The contiguous states and Hawaii are divided into rectangles 7.5' of latitude by 7.5' of longitude; thus, the maps are often called *7.5' quadrangles* or *7.5' quads*. They are also referred to as the *7.5' series*. Adjacent maps can be placed side by side to form a larger map. The entire country, minus Alaska, has been mapped at this scale, which requires more than 55,000 quads. Because Alaska is such a large and sparsely populated area, it has not been completely mapped at this scale, so coverage is a blend of scales. The basic scale is 1:63,360, or 1 inch to 1 mile, and the quads cover 15' of latitude, but more than 15' of longitude depending on location. Recall that meridians

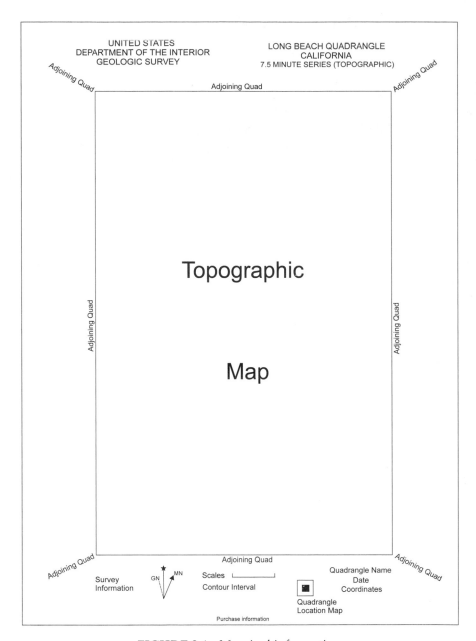

FIGURE 8.1. Marginal information.

converge at the poles, so the length of a degree of longitude varies, decreasing in linear distance as you move north along a meridian. There are 2,700 maps in the Alaska 15' series. Some populated areas have been mapped as 7.5' quads.

The first series of topographic maps to cover the United States was at a scale of 1:62,500 with 15' quadrangles. Maps are no longer being made at this scale, but the old maps are still available online as *historic quadrangles* (*http://nationalmap.gov/ historical*). These maps are quite useful in historical studies of an area.

The other current scales produced by USGS are 1:100,000 and 1:250,000 for complete coverage of the country. There are also some special topographic maps, such as a national park series. Figure 8.2 shows a comparison of the basic scales for topographic maps.

Projection and Grids

Until the 1950s, USGS used the polyconic projection for topographic maps, but it has been replaced with the transverse Mercator projection (see Chapter 5). Because of the small areas covered by topographic maps, the errors introduced by the projection are negligible, so distances and area can be measured with great accuracy.

The maps show tick-marks in the margins for the grids represented on the map, black marks represent latitude and longitude, and blue marks represent the Universal Transverse Mercator Grid. If a quad lies entirely within one grid zone for a State Plane Coordinate System, the grid lines are indicated near the southwest and northeast corners. The UTM grid is also shown in the margins. For states covered by the U.S. Public Land Survey System, townships and sections are shown.

Accuracy

Accuracy is important on topographic maps because of the ways in which they are used. It is vital that engineers and land-use planners have reliable maps for their work. Thus, the USGS has established *map accuracy standards*. The *horizontal accuracy* standard requires that 90% of the points tested must be accurate to within 1/50 of an inch (0.05 cm) on the map. At 1:24,000, 1/50 of an inch is 40 feet (12.2 m). The *vertical accuracy* standard requires that 90% of all points tested must be accurate to within one-half of the contour interval. On a map with a contour interval of 40 feet, all points will be accurate to 20 feet.

Symbols

The complete symbol sheet for topographic maps is lengthy (see Plate 8.1), but not all symbols are used on all maps. A complete list for the conventional (historic) maps can be found at *http://egsc.usgs.gov/isb/pubs/booklets/symbols/topomapsymbols. pdf*. There is some variation of the symbols used on topographic maps; some are used only on older maps, whereas others are found only on USGS/USDA Forest Service joint maps (see below). Because a variety of different kinds of features, both cultural and natural, are mapped, point, line, and area symbols are all utilized depending on the type of feature. Users of US Topo must also be aware that, because the base is an orthophoto, many of these colors and features are not shown on the current editions but will be in the future. The symbols used on US Topo maps can be found at *http:// nationalmap.gov/ustopo/images/US_Topo_Symbol_Sheet*.

Five colors are used for symbols on topographic maps: brown, blue, green, red, and black. On some older "provisional" maps, purple was used to represent new information that had not yet been field checked.

Brown is used for terrain features. The most obvious brown symbols are contour

FIGURE 8.2. Topographic map scales.

lines that join points of equal elevation. Contour lines never have gaps; that is, they are closed lines, and they never cross one another. The basic contour line is the *index contour*, which is a heavier line than the others and has its elevation marked on it. Index lines are usually every fifth contour. *Intermediate contour lines* are thinner and have no numbers. The elevation (vertical distance) between two adjacent intermediate contours is the *contour interval* of the map; this is usually specified in the map legend. In addition to these primary forms of the contour line, on some maps one will find dotted *supplemental contours* and hatched *depression contours*. Supplemental contours are used in areas where the terrain is so flat that no or few contours would occur at the specified contour interval. Thus, on a map with a 40-foot interval, a dotted brown line will represent 20 feet in the flat areas. Enclosed depressions are represented by depression contours indicating that the feature is lower than the surrounding terrain. These are found on volcanic craters and around sink holes. If the elevations have been estimated, a dashed *approximate* or *indefinite* contour line indicates that the lines might not meet accuracy standards (Figure 8.3).

Blue represents water features, whether natural or humanmade. Thus, rivers, canals, including irrigation and drainage canals, lakes, and reservoirs are shown in blue. Swamps and glaciers are also represented with blue symbols, as are *isobaths* (depth below the water). Isobaths are read in the same way as contour lines and allow the reader to determine the nature of the undersea floor.

The drainage patterns of an area provide a great deal of information about an

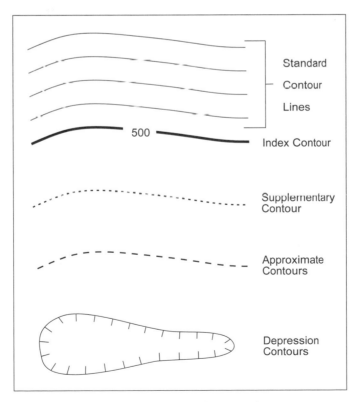

FIGURE 8.3. Kinds of contour lines.

area. Such patterns reflect something of the character of the underlying rock and the climate. In humid areas, streams are usually *perennial*; that is, they flow all year long. In arid areas some streams have an *intermittent* flow, which means that they are dry for a considerable portion of the year. On topographic maps perennial streams are shown with solid blue lines and intermittent streams are shown with a symbol made up of dashes and dots (see symbol sheet; Plate 8.1). When interpreting a map, you must be aware that perennial streams that originate in humid areas sometimes flow through deserts; the Nile in Egypt and the Colorado River in the Southwest United States are examples. These rivers are called *exotic* streams.

Different types of underlying rock structure result in various stream patterns (Figure 8.4). The most common pattern is tree-like and called *dendritic*; it forms on relatively uniform material. *Trellis* drainage is a rectangular or trellis-like pattern of relatively straight streams that meet at right angles. This is found where there is banding or structural weakness. *Radial* drainage occurs around a central mountain or dome, and the streams flow outward in all directions. *Centripetal* drainage is similar except that it is found on craters or depressions and the streams flow inward from all directions. A combination of radial and centripetal drainage usually indicates a volcanic cone. *Annular* drainage is similar to trellis drainage but is curved. It is found on structural domes. *Braided* streams occur when a stream is overloaded with sediment and drops some of the load, resulting in clogged channels.

Green symbolizes vegetation, both natural and planted. Green symbols aren't used on US Topo; the reader must interpret the kinds of vegetation from the imagery. Six different types of vegetation are shown on the conventional maps: woodland, scrub, mangrove, wooded marsh, orchards, and vineyards. The topographic maps of some countries show different crops, but on USGS maps crops are not shown because the crops planted in fields often vary from year to year. *Woodland*, symbolized in solid green, is an area of tree cover or brush that is at least 6 feet tall and dense

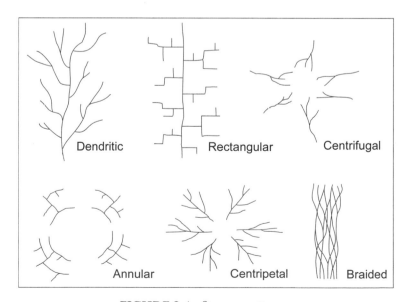

FIGURE 8.4. Stream patterns.

enough to provide cover for troops. This definition reflects the sometimes military uses of topographic maps. *Scrub,* symbolized by irregularly spaced green dots, is an area of low-growing perennial vegetation, such as mesquite or sagebrush that are both common in arid areas. *Mangrove* is a particular type of swamp; it is a dense growth of tropical, maritime trees with aerial roots. Mangrove is usually found in saline water in shallow bays and deltas, and along riverbanks. It is symbolized by a blue leaf pattern over a green background. *Wooded marsh* is symbolized by a blue "grass" pattern over a green background and is an area of normally wet land with tree cover. *Orchards,* symbolized by regularly spaced green dots, are plantings of evenly spaced trees that bear fruit or nuts. *Vineyards,* symbolized by regularly spaced smaller green dots, are plantings of grapevines that are supported and arranged in evenly spaced rows. For the purposes of mapping, other cultivated climbing plants, such as hops and berry vines, are shown with the same symbol.

Although we tend to think of topographic maps as primarily showing terrain, they also provide a wealth of cultural information. Because cultural features change more frequently than terrain features, one must be very aware of the date of the map; new roads may have been built and old ones removed, and housing developments may have been built on former orchard lands. Cultural features on USGS topographic maps (with the exception of those mentioned above) are shown in black and red. These cultural elements include transportation features, houses and landmark buildings, industrial complexes, and extractive activities, such as oil fields, pipelines, and mines. Also found are cemeteries, recreational areas, historical landmarks, boundaries, and names of features.

Transportation features include primarily roads and railroads. *Roads* are shown in five major classes: Class 1 are primary highways that are all-weather and hard surface; included here are interstates, federal routes, and primary state routes. The Class 1 symbol is a red line edged with solid black. For numbered highways, the highway symbol and number are also shown. Class 2 roads are secondary highways and are symbolized with a dashed red line edged with solid black. Light duty roads, Class 3, both paved or gravel, are shown with a variety of black symbols. These first three classes are always included on topographic maps. Classes 4 and 5 include unimproved roads that are passable in dry weather, as well as trails used for bridle and pack trails. Maintenance roads and four-wheel drive roads may be included and are identified by labels.

Railroads are shown in black, and the major division is standard gauge and narrow gauge. Gauge is the distance between the rails of the track; standard gauge is 4 feet 8.5 inches, and narrow gauge is any rail separation less than that. All mainline railroads are standard gauge, while narrow gauge is used mostly on private freight railroads. A single black line with "ties" indicates single track, whereas two parallel black lines with ties represent multiple tracks. A caution is in order here: Two lines are used whether there are two, three, or more tracks. Sidings and spurs are mapped accurately in length, but the position may be adjusted owing to the scale of the map. *Railroad yards* are mapped in outline based on the outer tracks, but the exact number of tracks within the yard cannot be determined, again owing to generalization for the scale of the map. If *airports* and *airfields* are large enough, their runways and terminals are shown on the map. Small landing fields or landing strips are represented by a symbol.

Buildings are the most common human-made features after roads. Buildings intended for housing or human activities such as residences, hotels, churches, schools, shops, factories, and most public functions are shown in solid black or with cross-hatching. Buildings that are not intended primarily for human activities, such as barns and warehouses, are shown by an open black square or a square with diagonal lines. If the building is large enough, its actual outline or "footprint" is shown. In highly urbanized or "built-up" areas, only landmark buildings are shown, and the built-up area is covered with a pink tint. This is called a *building omission area.* Landmark buildings are schools, churches, city halls, post offices, and the like; the street network is also shown. On US Topo, the individual buildings can be seen on the orthophoto; they are not symbolized. Thus, in urban areas all houses are visible, but schools, churches, and other buildings, though visible, are not named.

Included in *industrial and mining* are such things as refinery operations, large industries, open-pit mines, strip mines, and below-surface mines and their associated features, such as tank "farms," narrow-gauge railroads, mine tailings, and mine dumps. Extractive industries (mining and oil drilling) are easily recognized on topographic maps. Open-pit mines and quarries are most often depressions and when large enough are labeled. Underground mines have symbols for adits (entrances), shafts (vertical openings), and prospects (openings for mineral exploration). (See Plate 8.1 sheet.) Mine tailings and dumps are the discarded materials and are shown in brown; contours are used if the dump is large enough. Oil and gas wells are small open black circles labeled to show the type. Again, while these features can be seen on the orthophoto image, US Topo does not label them, and so the map reader must guess at identification. These features will probably be included in future versions.

Recreation areas range from city parks to national parks. These areas are labeled on the map with boundaries shown. Larger recreational areas may have campgrounds, hiking trails, and visitor centers. It is important to understand the differences between national parks, national monuments, and national forests. *National parks* and *national monuments* are protected areas that are under the auspices of the U.S. Department of the Interior; the National Park Service manages the parks and monuments. The difference between a national park and a national monument is that national parks are created by Congress and monuments can be established quickly by a president without the approval of Congress. Although the two are both protected areas, there are differences in funding. *National forests* are a part of the U.S. Department of Agriculture and have a different mission. The slogan "Land of Many Uses" is applied to the national forests; grazing, mining, and hunting may be permitted in national forests. More information on parks and monuments is found in Chapter 11, "Maps for Special Purposes."

Reading and Analyzing Contour Lines

When you first look at the contour lines on a topographic map, they appear to be a tangle of brown lines, but with experience you can visualize the landforms, the ups and downs of the terrain, the cliffs and valleys. With a little effort, you can quantify the steepness of an area, or draw a picture of the ups and downs of the trail you will hike or bike. You can determine if the slope of the land is too steep to plant row crops or if contouring or terracing is necessary. All of this work is done with contour lines, some basic arithmetic, and a few simple tools.

We learned earlier in this chapter that contour lines join points of equal elevation; that is, every point on a given contour line is the same elevation above sea level. The lines on a contour map are drawn at a specific interval that is shown on the legend. The *contour interval* is the *vertical* distance between two adjoining contour lines (see Figure 8.5). When we look at the map, we see that the lines vary in their distance from one another. Some areas have closely spaced lines, and others have lines that are quite far apart. This horizontal or map distance between lines is called the *spacing* of the contour lines. Figure 8.6 shows cross sections of three islands and the contour maps that could be drawn for each. Notice that where the islands are steep, the lines are close together; where the slope is more gentle, the lines are farther apart; where the slope is even, the contour spacing is even; and where the slopes are irregular, the contour spacing is also irregular. Contour lines also follow certain "rules." Contour lines never cross one another, and they always close on themselves either within or outside the map. This means that contour lines cannot start and end abruptly. When contour lines cross a stream valley, because they maintain the same elevation, the lines point upstream (Figure 8.7). This is the basic information needed to determine the nature of the terrain from a contour map. Obviously, the more practice you have, the more skilled you will become. Taking a topographic map into the real world and comparing it with the actual landscape also helps in understanding the relationship between contours and landforms.

Gradient and Slope

Sometimes, we want more than just the general idea of the nature of the terrain, we want to know, "How steep is that trail? When we want to determine the steepness of a trail or other linear feature, such as roads or rivers, we want to find the *gradient* (or

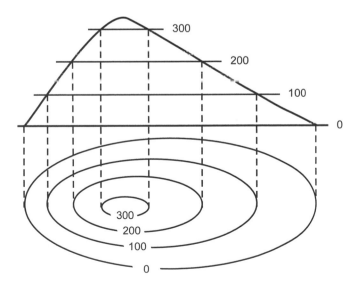

FIGURE 8.5. Contour interval and spacing.

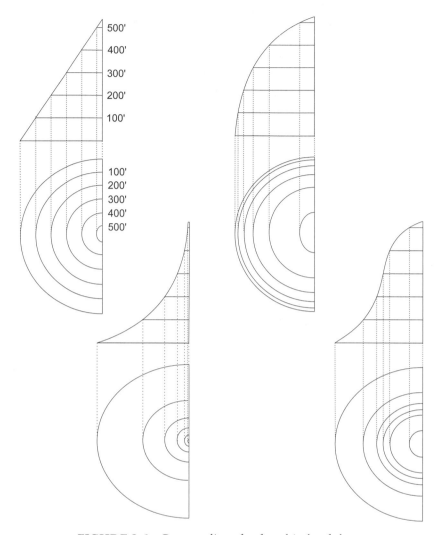

FIGURE 8.6. Contour lines for four kinds of slope.

grade) of the trail. The gradient is the inclination of a *linear feature* to the horizontal; it can be expressed as a degree, as *rise* over *run*, or more commonly as a percentage. The rise is the difference in elevation between the start of the trail segment, and the run is the *map distance* of the segment. Notice that because the trail is inclined, the map distance will be somewhat different from the actual distance of the trail. The percent of the gradient is found by the following formula and in Figure 8.8.

$$\text{Rise/Run} \times 100\% = \% \text{ gradient}$$

Thus, if the difference in elevation is 200 feet and the length of a trail is 2,000 feet, the gradient is 10%. This is the average gradient and assumes a straight line. If the run has twists and turns, we can determine its length either by marking the entire length along the edge of a piece of paper and comparing it to the graphic scale or by

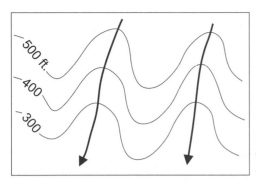

FIGURE 8.7. Contour lines "point" upstream when they cross a stream valley. The direction of the stream flow is indicated by arrows.

using dividers to mark off scale distance along the trail. We must be aware that the gradient along the trail may vary: There will be steep segments and gentle segments. Thus, when we use the entire length of the trail, we will get only an average gradient and will miss possibly important parts of the picture. This problem can be solved by calculating the gradient for trail or road segments.

The gradient is important not just for hiking and biking a trail, but for roads and railroads. For example, the maximum grade allowed for interstate highways in hilly areas is 6%; other roads may have steeper grades. It is common to see highway signs warning of steep grades with the percent of grade shown (Figure 8.9). In Figure 8.9, the decrease in elevation is 7 feet for every 100 feet of horizontal distance. The maximum grade for most standard gauge railroads is 3%, but most are less. It is important to understand that a 100% gradient is *not* a vertical cliff. A 100% gradient has a slope angle of 45°. Figure 8.10 shows slope angles and percentages of slope.

Slope is the inclination of a *surface*, such as a field, compared to the horizontal. We might want to determine the slope of a hillside to assess its stability or its potential for farming. We determine slope in much the same way as gradient. Again, we must remember that the slope may not be uniform, and we may want to determine individual sections of the slope. The first step is to determine the rise for the area; second, a slope line is selected for the run measurement. The slope line should cross the majority of the contours at a right angle. These values are put into the formula, and the slope along the line is calculated. In general, nonterraced or contoured agriculture is carried out on slopes of 8% or less.

A slope map of an area can also be made in order to see the variation in slopes. This is accomplished in the following manner. Since we can determine the slope between two adjoining contour lines, if we make a scale that shows contour spacing (number of contours per inch) and percent of slope, such as 0–10, 10–20, or 20–30,

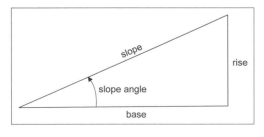

FIGURE 8.8. Determining gradient or slope. The base is the map distance, and the slope is the actual distance.

FIGURE 8.9. Percent of grade on a highway (photograph).

that scale can be moved around on the map to identify different slopes and those areas are shaded accordingly. Usually, a predetermined set of slopes is used.

One of the most useful tools for understanding terrain on topographic maps is the *profile*. A profile is a cross section of the land drawn along a given line, and it allows us to view the ups and downs of the terrain. Profiles are used by engineers and planners, but also by fans of outdoor recreation—hiking, biking, skiing. Figure 8.11 shows the steps in drawing a simple profile. Two scales are involved in profile drawing: *horizontal scale* and *vertical scale*. The horizontal scale is the same as the scale of the map. The map scale in the figure is 1:120,000. The vertical scale marks the elevation of the points along the line, and it may be *unexaggerated* or *exaggerated*. Unexaggerated profiles have a vertical scale that is the same as the horizontal scale. In Figure 8.11, 0.5 inch represents 5,000 feet. If the contour interval is 1,000 feet, then the vertical scale will show 1,000 feet as 0.10 inch. If the total elevation is

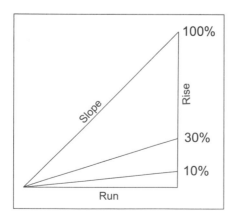

FIGURE 8.10. Percent of slope (diagram).

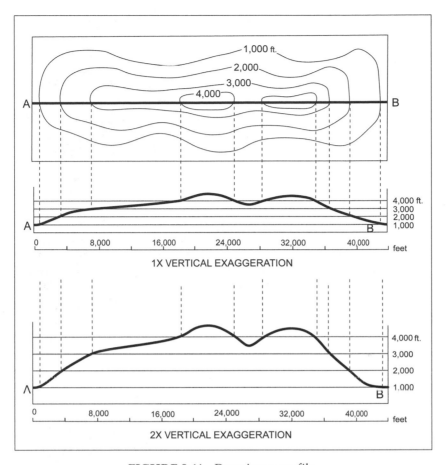

FIGURE 8.11. Drawing a profile.

5,000 feet, the profile will only be 0.5 inch high; this is not big enough to show much detail. For that reason, profiles are often *exaggerated*. On exaggerated profiles, the vertical scale is larger than the horizontal. For our example, the horizontal scale is 1 inch represents 10,000 feet, and the vertical scale is 1 inch represents 5,000 feet; now the nature of the terrain is more easily seen. More exaggeration is needed in relatively flat areas than in steep areas.

Profiles can also be drawn along roads and trails, although this is somewhat more cumbersome. Here the length of the trail serves as the profile line, and the profile is made up of a series of segments (Figure 8.12). A number of software and online programs can create profiles, such as *Topo!* and *Map My Run*, but it is not always possible to control the amount of exaggeration with these programs. While the software can be time saving, you may not have access to it when in the field. A basic knowledge of profile drawing and exaggeration will help in understanding the profiles that result from these programs and also prevent the hiker, biker, or runner from being intimidated by a relatively flat area that appears very steep because of the amount of exaggeration or, conversely, from assuming an area is flatter than it is. We can see from Figure 8.12 that the gradient differs along the line.

Land Survey Systems

We have looked at the various types of cadastral systems in Chapter 4. The patterns formed by these systems can be seen on topographic maps and give clues as to when and by whom an area was settled. PLSS patterns are especially obvious in the Midwest of the United States where the roads follow the township and range lines. The unsystematic surveys found in much of the eastern part of the United States gives a "crazy-quilt" pattern to roads and property lines, and the long, narrow properties of the French Long Lots tell us that an area was settled by the French.

Other Topographic Maps

While we most associate the standard USGS maps with topographic maps, some commercial organizations, notably National Geographic Society, Garmin, and Delorme mapping, produce selected topographic maps of the United States. National Geographic makes topographic maps of trails, and Delorme makes topographic atlases of each state, with maps at a scale of 1:100,000. Both Delorme and National Geographic sell topographic maps of selected areas on DVDs. These maps allow profile drawing and slope and gradient calculation.

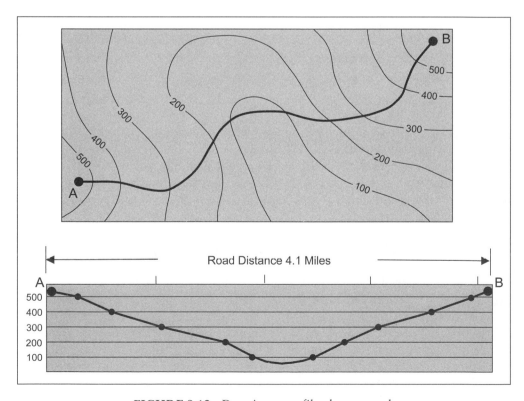

FIGURE 8.12. Drawing a profile along a road.

GPS and Topographic Maps

In the past 15 years, hundreds of books have been written about GPS, giving the impression that GPS has replaced topographic maps, that topographic maps are obsolete. Actually, topographic maps on paper or computer and GPS complement each other and are used in conjunction. Because GPS can fail or satellite transmissions can be blocked in certain conditions, such as deep canyons, as was noted in Chapter 7, hikers are advised to carry a paper map and conventional compass. Certainly, with GPS units that show only position, direction, and altitude, a topographic map is a necessity. Most GPS units used for hiking and other outdoors activity now have maps built in; some include topographic maps with contours and relief shading. These units also have the capability of drawing profiles of the route traveled and calculating the gradient.

What we must keep in mind with both maps downloaded from DVD and those on GPS is that, although the maps and profiles are displayed on a screen, an understanding of how to read those contours and of what the profiles and gradients show and mean, as well as how to interpret them, is essentially the same as for paper maps.

Interpretation

Each topographic map has thousands of bits of information, and analyzing and interpreting any such map can seem a daunting job. Table 8.1 provides some guidelines for interpretation. In Chapter 12, we will look at a topographic map and analyze it.

FURTHER READING

Evans, R. T., and H. M. Frye. (2009). *History of the Topographic Branch (Division)* (U.S. Geological Survey Circular 1341). Available at *http://pubs.usgs.gov/circ/1341*.

Hinch, Stephen W. (2007). *Outdoor Navigation with GPS*. Berkeley, CA: Wilderness Press.

Kjellström, Björn. (2010). *Be Expert with Map and Compass* (3rd ed.). Hoboken, NJ: Wiley.

Letham, Lawrence. (2008). *GPS Made Easy* (5th ed.). Seattle, WA: The Mountaineers.

Thompson, Morris M. (1988). *Maps for America* (3rd ed.). Washington, DC: USGS.

U.S. Department of the Army. (2001). *Map Reading and Land Navigation, FM 3-25.26*. Washington, DC: Author.

RESOURCES

USGS Store
  http://store.usgs.gov
Historical topographic maps
  http://nationalmap.gov/historical
US Topo
  http://nationalmap.gov/ustopo
US Topo Users Guide 16-page introduction
  http://nationalmap.usgs.gov/ustopo/quickstart.pdf

TABLE 8.1. Topographic Map Interpretation

A topographic map carries a wealth of information. To a large degree, the interpretation of such a map depends on the ability to distinguish many overlapping patterns, one from one another. You may proceed toward an interpretation by considering some or all of the following items and adding others as relevant to the particular area.

1. *Introductory information:* Location, both absolute and relative; area (size of subject area); primary land/water relationship; source of information and degree of accuracy (contour interval and scale).

2. *Terrain:* Highest and lowest points; local relief (difference between high point and local low point); descriptive characterization (plain, plateau, etc.); overall similarities versus sectional differences; regional slope (direction and amount).

3. *Drainage:* Pattern (gross and detailed); major streams; gradient; unusual features (waterfalls, etc.). Streams seek out minute differences in the rocks in their erosive work; consequently, there is no better source of information regarding the geology of an area than the pattern of the streams and the shape of the landforms they produce.

4. *Cultures:* Transportation patterns, road, rail, and other; transportation features (junctions, rail yards, ferries, dead ends, class of rail or road). Settlement sizes and types, plus implications of those shown only partially at the edge of sheets; economic information, mines, orchards, ports, tramways, and so on.

5. *Occupancy:* Population density and pattern of distribution; economic production, problems of transportation construction, water problems in terms of adequacy or excess related to possible uses (domestic, industrial, navigation, drainage); isolation or integration (gross and detailed); institutions (schools, hospitals, shopping); size of farms, types of crops.

The foregoing is *not* a checklist, nor is it an order of presentation, nor is it complete. It is a list of suggestions only, which should stimulate the discovery of many bits of information about the area. The actual presentation should describe the area of the map, the land, and the human occupancy of it.

Note. Courtesy of Gerard Foster.

Brief history of the USGS
> *http://nationalmap.gov/ustopo/125history.html*

Topographic Map Symbols
> *http://egsc.usgs.gov/isb/pubs/booklets/symbols/topomapsymbols.pdf*

US Topo symbols
> *http://nationalmap.gov/ustopo/images/US_Topo_Symbol_Sheet*

The National Map
> *http://nationalmap.gov*

Map Scales Fact Sheet
> *http://egsc.usgs.gov/isb/pubs/factsheets/fs01502.html*

CHAPTER 9

Thematic Maps

> Maps, like stories, have a main theme, point of view, plot, and style.
>
> —*Mapping* (p. 175)

Thematic maps are maps that show one or more *themes*, such as population, land use, vegetation, and voting patterns. They have also been called *special-purpose maps* and *statistical maps*. However, these terms are generally inappropriate because "special purpose" is often used for large scale maps used in particular fields, such as geologic maps. The term *statistical map* is also inappropriate because not all thematic maps are statistical, that is, quantitative.

Thematic maps have two parts: the base map layer that consists of such elements as boundaries, perhaps basic hydrography, major cities, and other reference information; and the thematic or subject information layer. Thematic maps range from quite simple, such as showing point locations of towns or vegetation patterns, to quite complex, such as showing interrelationships between themes. Far more has been written about making thematic maps than reading and interpreting them, and yet almost every day we encounter thematic maps; they are in newspapers, magazines, textbooks, and atlases. Thematic maps are sometimes created with a bias and may be persuasive or even propagandistic, but the majority are not deliberately biased.

Thematic maps had their origins in the late 17th and early 18th centuries. Edmond Halley, of Halley's Comet fame, is usually given the credit for their "invention" with his maps of winds and magnetic declination. Thematic maps became popular with the rise of statistics and national censuses in the late 18th century when statistical data became available for mapping.

Thematic maps are usually divided into qualitative and quantitative. *Qualitative thematic maps* show location patterns and some "quality," such as climate, vegetation type, or land-use types. No percentages, ratios, or other quantities are involved. *Quantitative thematic maps* show locations, categories (qualities), and amounts, such as size of towns, percent of ethnic population, amount of rainfall, or temperatures. All of the symbol types discussed in Chapter 3 are used for thematic maps—points, lines, areas, volumes—and for animated maps, time.

QUALITATIVE THEMATIC MAPS

Qualitative maps show the location of the feature(s) plus its nature. Thus, on a map with point symbols, the point might show the location of buildings and some non-numerical quality, such as their use. Linear symbols show location and direction as well as some quality such as type of road—highway, secondary road, or unpaved. Areal symbols show the location and extent of the feature (e.g., urbanized area, forest, or agriculture) and some information about type. Thus, for forests the map might show deciduous and coniferous trees; for agriculture it might show ranching, crops, dairy, subsistence, or grains. Although the size of the area represented could be construed as quantitative, for these maps it is the nature of the feature that is important.

Because lines and colors are used to separate the different groupings, such as coniferous and deciduous, the reader may get the idea that there are sharp distinctions between the areas, but in reality, such distinctions rarely occur. The "line" that separates the two types is not a sharp division in the real world; along that line there is probably a mixture of coniferous and deciduous trees. The uniform colors or tones also give the impression of total uniformity, for example, that in an area shown as deciduous, there are no coniferous trees. Sharp divisions and areas of complete homogeneity do not usually occur in the natural worlds. How then do we interpret such maps?

First, we must remember that these maps are usually small scale; in fact, in atlases, the thematic maps are some of the smallest scale maps, as small as 1:100,000,000. Because of these scale limitations, these maps are highly generalized. Thus, we must realize that in the forest example the distribution shown is the *predominant* type of forest in the area mapped. Since sharp divisions between phenomena rarely occur in nature, the lines cannot be precisely drawn and easily visible in the real world; the lines can be treated as transition zones. Cartographers have been challenged by this problem since the beginning of thematic maps and have tried to solve it in a variety of ways—by blending colors or by interdigitation—but lines are still most common (Figure 9.1).

We look at patterns of distribution, and we might correlate the map with one or more maps that show other features to help explain the patterns. This issue will be covered in more detail in Chapter 12.

QUANTITATIVE THEMATIC MAPS

Point Symbols

Like qualitative maps, quantitative thematic maps may use point, line, or areal symbols. The point symbols are geometric figures, such as circles, squares, and triangles,

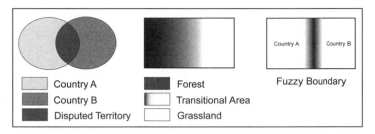

FIGURE 9.1. Overlapping symbols. These symbols are used to show transitions, areas of dispute, and indeterminate boundaries.

although they are usually called dot maps, graduated circle maps, or proportional symbol maps.

The *simple dot distribution map* shows distributions generally by means of small round dots, each with an assigned value, that are placed in the center of the density of distribution for the given area. Theoretically, it should be possible to count the dots on a map and arrive at total numbers, but in practice only rarely can this actually be done; on most dot maps, the dots coalesce and form a solid mass in the densest areas of concentration. In fact, it isn't desirable to count the dots. Most such maps list the source of the original data, and those tables are much more useful for determining total numbers than the map. On some maps the totals are given. Where dot maps excel is in portraying the *pattern of distribution* of the subject (Figure 9.2).

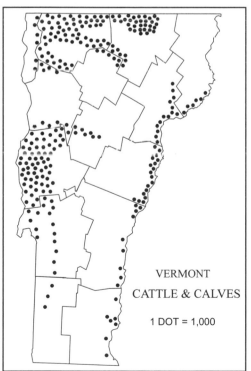

FIGURE 9.2. Simple dot maps show quantity and the nature of the distribution.

A variation of the dot map is the *dot density map*. Such maps look much the same as simple dot maps, but they are not made or interpreted the same way. To make a dot density map, the cartographer again assigns a value to each dot, but then, instead of placing dots in the center of distribution, the dots are placed randomly within the area. This provides a picture of the density of the product but not its distribution pattern. In the past 20 years, dot density maps have become more common than dot distribution maps because they are easier and quicker to create with a computer; the cartographer does not have to determine where the product is found within the area (Figure 9.3).

In spite of their apparent simplicity, both kinds of dot map can be misinterpreted. As an example, if one dot represents 100 head of cattle and a county has 100 head, it is entitled to one dot, which on the simple dot map is placed where the majority of cattle are found. It would be a mistake to assume that there are no cattle anywhere else in the county. It is also common for an area to have too few cattle (or other product) for even one dot (usually less than half of the dot value), and the county receives no dots. Again, it is a mistake to assume that the county has no cattle. When interpreting dot density maps, it is a mistake to assume that the pattern of dots represents the distribution pattern of the product, since these maps only show relative density, not locations. It is, therefore, important to understand which kind of dot map is being used. A good map will indicate if it is dot density.

The other type of point symbol is the *proportional point symbol*, often called the *graduated circle* although it may have shapes other than circles. In the simplest form, the size of the circle represents totals; in Figure 9.4 the circles represent city popula-

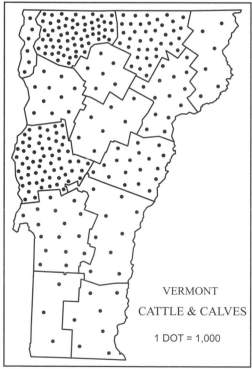

VERMONT

CATTLE & CALVES

1 DOT = 1,000

FIGURE 9.3. Dot density maps do not show actual locations.

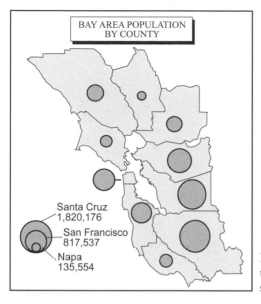

FIGURE 9.1. Proportional circle. The area of the circle is proportional to the amount represented.

tions. These symbols are constructed so that the *area*, not the radius or diameter of a given circle, is proportional to the amount represented. In Figure 9.5, circle A has a radius two times that of circle B, but the area and value are four times larger. A different circle size is drawn for each quantity. These symbols are shown in the legend with a scale. A second type of graduated circle divides the data into several categories, and each category has a different-size circle. These are somewhat easier to read and understand (Figure 9.6).

Proportional symbols may apply to point locations such as when city populations

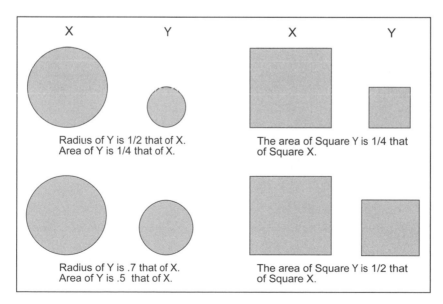

FIGURE 9.5. Proportional shape sizes.

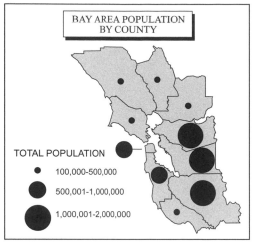

FIGURE 9.6. Range graded circles. Values are placed in categories.

are shown, or they may apply to regions. Although the symbols are usually circles, this is not always the case; squares, triangles, stars, and any other geometric figure may also be used. Often the figures are subdivided to show the parts that make up the whole. Thus, the total number of a state's registered voters might be shown with a circle subdivided according to Republican or Democrat (Figure 9.7). These segmented circles are commonly called *pie charts* or *pie graphs*.

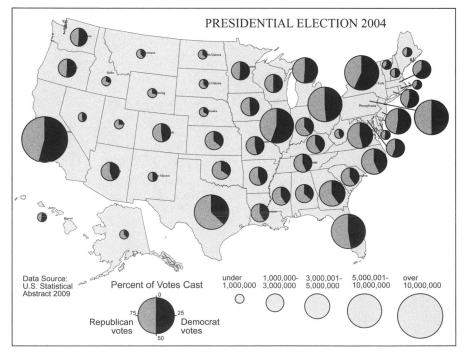

FIGURE 9.7. Pie charts.

Line Symbols

The *flow map* is a simple type of quantitative line symbol that uses the width of the line to represent the amount. A common use of flow maps is to show volume of traffic with or without an indication of the direction of flow. Flow maps are also used to show immigration to and emigration from a country as well as imports and exports (Figure 9.8).

Mapping Volumes with Lines

Plates 12.2–12.6 in Chapter 12 are a series of maps of Australia. Those showing temperature and precipitation are *isarithmic* maps. *Isarithms* are lines that join all points having the same value. The contour line, which was discussed in detail with topographic maps in Chapter 8, joins all points having the same elevation above sea level and is the most familiar form of isarithmic line. The same basic rules apply when reading these lines as when reading contours. In fact, the distributions may be viewed as an imaginary three-dimensional surface with hills and valleys of population, rainfall, or temperature. This imaginary surface is called the *statistical surface*.

Isarithms are drawn at specific values such as 10, 20, 30, 40 inches or 0, 20, 40 degrees. The difference between lines is called the *isarithmic interval,* and it may be even as in the examples above or uneven as is common with maps of population, where the lines might be 2, 25, 150, 250, and over 250 people per square mile or square kilometer. Uneven intervals can make interpretation more difficult because it distorts the impression of steepness or flatness (Figure 9.9). The *rate of change*, which can be thought of as analogous to slope or gradient of terrain, can be important when interpreting isarithmic maps; the closer together the isarithms, the more rapid the rate of change, that is, the more rapidly the values are changing. As is the case with

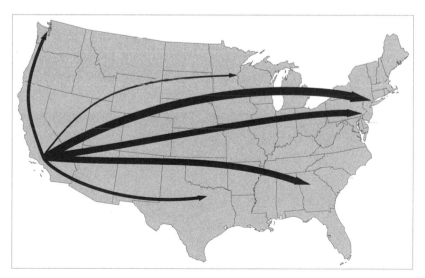

FIGURE 9.8. Flow line map. The width of the arrows is proportional to the value of the product.

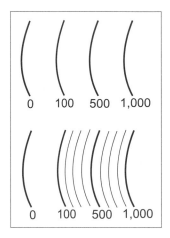

FIGURE 9.9. Uneven intervals can mislead.

contour maps, we can construct profiles of isarithmic maps that tell us how rapidly amounts are increasing or if there is an even rate of change.

The cartographer creates an isarithmic map by establishing *data points*. These can be weather stations for climatic data or the center of census tracts for population. He/she determines an isarithmic interval and draws the isarithmic lines by connecting values (Figure 9.10).

It is common for color to be added between isarithmic lines to make it easier to distinguish the lines—a black-and-white isarithmic map is hard to read. This is *layer tinting* as discussed in Chapter 3. The problem with such tinting is that the reader may assume that the colors have some special significance and see the statistical surface as a series of layers rather than a sloping surface.

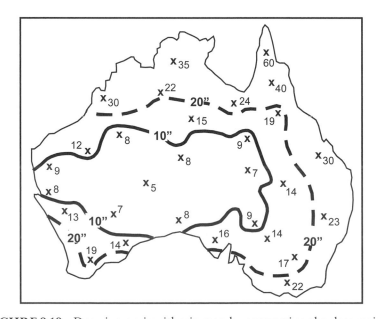

FIGURE 9.10. Drawing an isarithmic map by connecting the data points.

There are two types of isarithmic map—those that show data that can occur at actual points, such as rainfall, and those whose data do not exist at actual points, such as population per square mile. The first are sometimes called *isometric lines*, and the second are *isopleths*. Isopleths do not show exact values, but rather derived values such as population per square kilometer, percentages, and ratios that apply to an area.

The amounts of rainfall, or the temperature of a place on isarithmic maps is determined by *interpolation*, which is a way of "reading between the lines." If the point of interest is midway between the 20- and 40-inch isohyets (lines of equal rainfall), the rainfall can be estimated as 30 inches. If the point had been 3/4 of the way from 20 to 40, it might have been about 35 inches. It is important to keep in mind that these are estimates.

Isopleths present some problems in interpretation. Because the quantities shown are always derived quantities, ratios, and percentages, such as population per square mile or percent unemployed for an area, they cannot exist at a point, and interpolation is quite general. It is best on these maps to look at rates of change.

Area Symbols: Choropleth Maps

Choropleth maps are quantitative maps in which an area, such as a county, census tract, state, country, or other region, is colored or shaded according to the value of the item being mapped. An example is population per square mile by state. These maps are quick and comparatively easy to create, especially with modern computerized methods (GIS), but there are pitfalls. While the subjects for such maps are numerous, for them to be useful and meaningful, the statistics must be as ratios or percentages, not absolute numbers. The interpretation problem that occurs is shown in Figure 9.11 and was previously described in Chapter 3. Here we use a real-world situation. The size of states of the United States varies greatly; thus, if the total state population is mapped, since they have similar populations, Vermont and Alaska will be shaded alike, giving the impression that they are the same. If they are shaded according to population per square mile, however, Vermont has 67.9 people per square mile and Alaska only 1.2. A map of density is a more meaningful and accurate portrayal of the population. In Figure 9.11 the two states are drawn in their correct size relationship to one another. When reading choropleth maps, beware of those that utilize absolute numbers unless all of the areas are the same size.

The lines surrounding areas on choropleth maps have no meaning or value, unlike those on isarithmic maps; they are simply the boundaries of the enumeration areas. The value is applied to the entire area regardless of any variations that might occur within the area. This can result in sharp divisions between adjacent areas where actually there may be none, or in empty areas being lumped with the whole area. The statistical surface of choropleth maps can be visualized as a series of blocks whose bases represent the enumeration areas and whose heights represent the values. There are no smooth gradations on choropleth maps as there are with isarithmic maps.

A variation of the choropleth map is the *dasymetric map*. Dasymetric maps attempt to show variations within an area. Using our county example from Chapter 3, for a 100-square-mile county with 5,000 people, the population might be distrib-

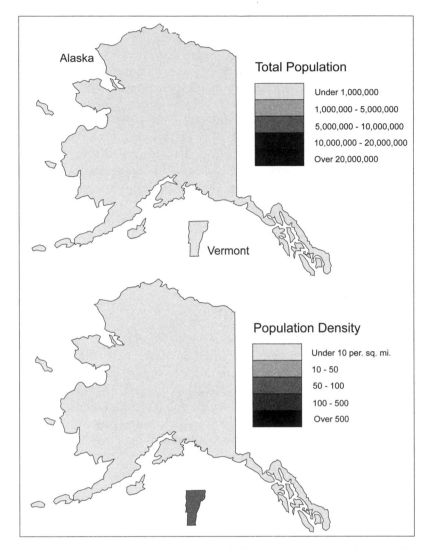

FIGURE 9.11. Alaska and Vermont have very different sizes but similar populations; thus, showing the total population instead of a ratio is misleading.

uted with 3,000 people in a town that is 10 square miles, zero people in a swamp that is 15 square miles, and 2,000 people in the remaining 25 square miles. On a simple choropleth, the entire county would be shown as 100 people per square mile, which doesn't correctly represent the area. To make a dasymetric map, the cartographer draws lines that represent the zones of rapid change, computes their densities, and shades them appropriately. For this county, the town has 300 people per square mile, the swamp 0 people per square mile, and the remaining area 80 people per square mile. Like the simple choropleth, the lines have no value, but here they represent the lines of rapid change, not political boundaries (Figure 9.12). A real-world example of the need for dasymetric maps is Egypt's population. For the entire country, the population is about 81 people per square kilometer, and that is all that a simple cho-

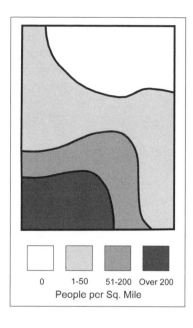

0   1-50   51-200   Over 200
People per Sq. Mile

FIGURE 9.12. The lines on dasymetric maps represent lines of rapid change; they have no value.

ropleth map would show. But since the majority of Egypt's population live along the Nile where the density is up to 5,000 people per square kilometer and few live in the adjoining deserts, which have less than 5 people per square kilometer, a more accurate representation is a dasymetric map that shows this.

CARTOGRAMS

Cartograms are a type of map on which size or distance is scaled to a variable other than earth size or distance units. The map may be scaled to population, time, or cost among other themes. Cartograms of various types are becoming increasingly popular because computers have made them easier to construct. We see cartograms with newspaper articles, online, in magazines, and in atlases. Such maps have existed since the 19th century, when statistical maps became popular and some were designed to help children understand geography.

Several kinds of cartograms are commonly used; the kinds most often seen are variations of the *value-by-area* cartogram on which the size of areas varies according to the value represented rather than the actual geographic area. Thus, the size may be based on population, income, electoral votes, average age, or crime rate for *enumeration areas*. Within this type are several subcategories (see below). A second major type of cartogram uses a time scale rather than a distance scale; thus a road's length is shown according to the time between two places rather than the number of miles separating them. These are called variously *distance cartograms* or *linear cartograms*. Probably the most famous cartogram is the diagrammatic map of the London Underground that can be viewed online (see Resources at the end of the chapter). This type of cartogram has been copied by transit agencies throughout the world (Plate 9.1).

Value-by-Area Cartograms

Contiguous value-by-area cartograms try to maintain borders between areas, but shapes are distorted. The rectangular form of these cartograms reduces every area to a square or a rectangle (Figure 9.13). A newer form of contiguous cartogram is the *topologically correct cartogram* (Plate 9.2). On these cartograms, the outlines and locations are highly distorted, and these are strictly computer created. Other variations replace the enumeration areas with uniform, abstract shapes, such as squares or circles (Figure 9.14). *Noncontiguous value-by-area-cartograms* show shapes correctly and either reduce or enlarge the enumeration areas according to the amount represented. Usually some effort is made to put the various areas in roughly their correct relative positions, but they are separated by empty spaces. Often a true outline of the area is placed around the enumeration areas to give the reader a frame of reference (Figure 9.15).

Some value-by-area cartograms represent two or more variables, such as population and income. One variable is shown as a cartogram, and the second variable is shown as a shade or color as on a choropleth map (Figure 9.16). This is often done on rectangular cartograms and used in many atlases.

Distance-by-Time Cartograms

Distance-by-time cartograms are also called *linear cartograms* and vary map distances according to the time needed to travel the real-world distance. Some preliter-

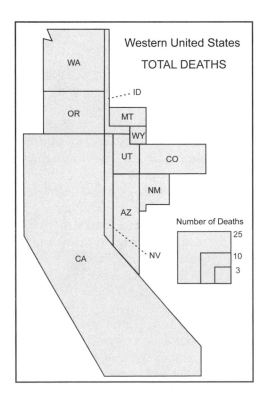

FIGURE 9.13. Rectangular cartogram.

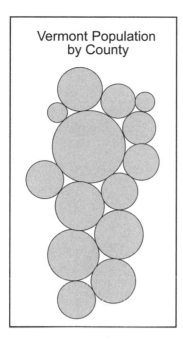

FIGURE 9.14. Dorling cartogram represents counties with circles.

ate peoples normally made their maps according to a time scale because the time required to make a journey was more important than the actual distance. Today, we often express distances as time, especially when describing freeways or expressways; two places may be the same number of miles from our starting point, but because of traffic conditions, terrain, traffic signals, and the nature of the road, one might be 30 minutes away and the other 1 hour away (Figure 9.17).

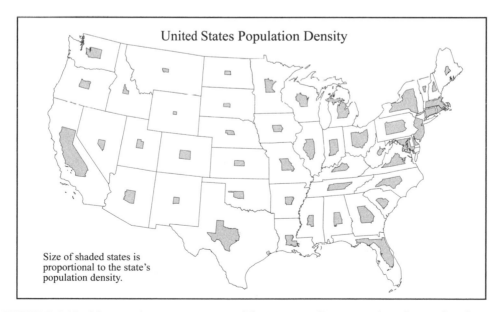

FIGURE 9.15. Noncontiguous cartogram. The state outlines are often shown for clarity.

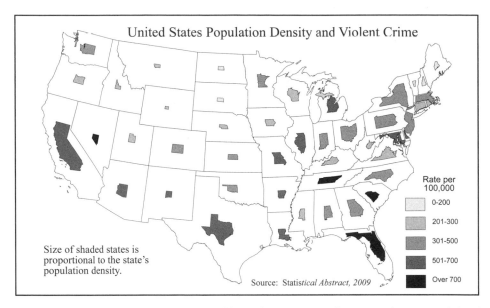

FIGURE 9.16. Cartogram with two variables. The state sizes show population, and the shading represents the crime rate.

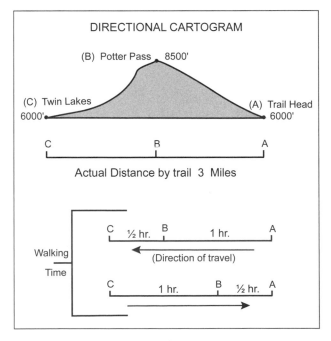

FIGURE 9.17. Linear cartogram.

Two kinds of cartogram are used to represent time. The first is linear and represents time from point to point; the other shows time as concentric circles from a center point (Figures 9.18).

How does one read cartograms? First, we must remember that an unstated goal of many cartograms is visual impact. Actual spatial relationships are distorted, but there is little generalization of data. There is not much standardization of cartograms. On some, a legend shows the value by size of squares; in others, simply the visual impression of larger or smaller suffices. A time scale is usually shown for distance cartograms. But overall, cartograms are not designed to provide detailed information.

FURTHER READING

Monmonier, Mark. (1993). *Mapping It Out: Expository Cartography for the Humanities and Social Sciences.* Chicago: University of Chicago Press.

Monmonier, Mark. (1996). *How to Lie with Maps* (2nd ed.). Chicago: University of Chicago Press.

Ovenden, Mark. (2007). *Transit Maps of the World.* New York: Penguin.

Robinson, Arthur H. (1982). *Early Thematic Mapping in the History of Cartography.* Chicago: University of Chicago Press.

Schulten, Susan. (2012). *Mapping the Nation: History and Cartography in Nineteenth-Century America.* Chicago: University of Chicago Press.

Tyner, Judith. (2012). *Principles of Map Design.* New York: Guilford Press.

FIGURE 9.18. Radial time cartogram.

RESOURCES

Cartogram Home
   www.ncgia.ucsb.edu/projects/Cartogram_Central/index.html
Map of the London Underground
   www.tfl.gov.uk/assets/downloads/standard-tube-map.pdf
Worldmapper
   www.worldmapper.org

CHAPTER 10

Maps for Navigation

> Mapping might well be called the science of
> whereabouts.
>
> —*Mapping* (p. 3)

Maps for navigation or way-finding are probably one of the oldest kinds of maps.
Almost every type of transportation has a map for following its routes. They are
all designed to serve a specific purpose, that is, finding and following a route, and
as such, information considered extraneous to the purpose is not included on such
maps. Thus, nautical charts give little, if any, information about features on land,
aeronautical charts are not concerned with ocean depths, and automobile road maps
focus on the features of most interest to a driver.

NAUTICAL CHARTS

The oldest of the navigation maps are nautical charts that "depict the nature and
shape of the coast, water depths, and general topography of the ocean floor; locations
of navigational danger; the rise and fall of tides; and locations of human-made aids
to navigation" (*http://celebrating200years.noaa.gov/foundations/nautical_charts*).

Cultures that travel by water have always had some form of guide, whether it is
a list of directions or a series of maps. In the 13th century, after the invention of the
mariners' compass, *Portolan charts*, which show a complex network of wind roses,
were used for sailing in the Mediterranean (Plate 2.3). The Marshall Islanders of the
Pacific created three navigational chart series made of palm ribs and shells. The series
included local, regional, and teaching charts; the palm ribs indicated wave directions

(Figure 10.1). While these are the best known, other island societies created navigational aids. The United States has a 200-year history of nautical charting dating to the formation of the Coast Survey by President Jefferson in 1807.

U.S. nautical charts are now created by the Coast Survey Division of the National Oceanic and Atmospheric Administration (NOAA) and are available in a variety of forms: paper maps, print-on-demand charts, charts in PDF format, and electronic versions. However, as of April 13, 2014, paper charts will no longer be available through NOAA. Electronic navigational charts are available that display on computers, mobile apps, and electronic display systems. For those who want a paper chart, some companies have a contract for print-on-demand charts, and selected charts are available in a printable PDF format (Figure 10.2).

The complete list of symbols and abbreviations for nautical charts is published in a booklet titled *Chart #1: United States of America, Nautical Chart Symbols and Abbreviations,* available online at no cost at *www.nauticalcharts.noaa.gov/mcd/ chart1/ChartNo1.pdf.*

The types of symbols used on nautical charts fall into several different categories. The first of these is the nature of the coast. Surveyed coasts are shown by a solid line, and unsurveyed coasts by dashed lines. Cliffs, rocky coasts, sand dunes, mud, sand, gravel, rubble, and stony shores are also shown. As might be expected, harbors and ports are shown in detail. Very few land features are shown, and these are usually landmarks that can be seen from the sea. Lights, beacons, and buoys, which are of importance to the sailor, are marked, and the color and type of light and buoy are shown. Underwater hazards, such as rocks, reefs, wrecks, masses of sea weed, and pilings, are important dangers and are symbolized.

The numbers that are seen in the water indicate *depth soundings.* Soundings are the equivalent to spot elevations on a topographic map and show the depth of the water at that point. The map also shows *isobaths,* lines of equal depth of water, which may be expressed in feet or fathoms (1 fathom = 6 ft). The compass rose is a

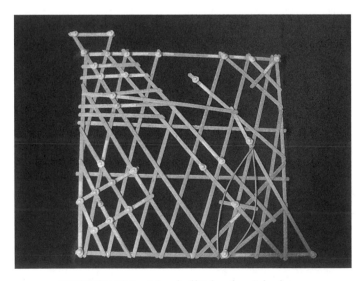

FIGURE 10.1. Marshall Islands stick chart.

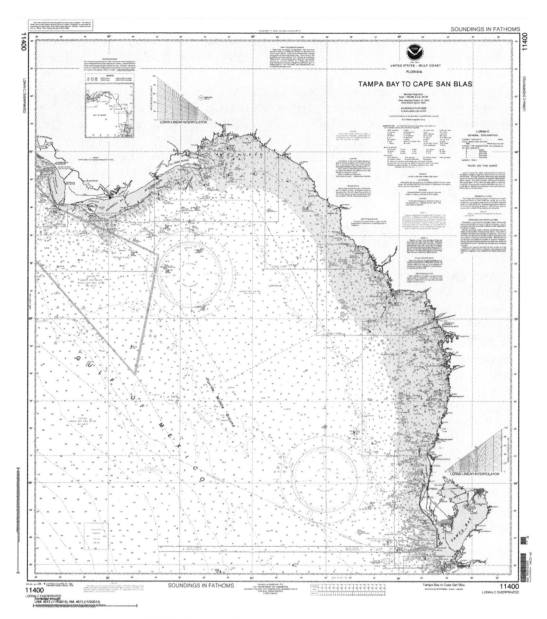

FIGURE 10.2. Nautical chart in PDF format.

striking feature of nautical charts. Both true north and magnetic north are marked, along with degrees for both true and magnetic north. Distance is usually expressed in nautical miles (6,076.1 ft), and speed is in nautical miles (knots) per hour.

AERONAUTICAL CHARTS

Shortly after the Wright brothers made their first flight, flying became popular. There were no specific maps for air navigation in those early days, nor were there instruments that would aid in flying in low visibility. Pilots followed roads and railroads for their routes, sometimes called "seat of the pants flying." By the 1910s, the first charts specifically designed for air navigation were made, and by World War II, millions of charts were produced.

An aeronautical chart is designed specifically for air navigation. Since it is a special-purpose map, it shows primarily features that are of importance to the pilot or air navigator flying by visual flight rules (VFRs) or instrument flight rules (IFRs). In the United States these charts are provided by the Federal Aviation Agency (FAA). There are basically two types of aeronautical charts: *instrument navigation charts* and *visual navigation charts*. Instrument charts aid the pilot who is flying and landing under IFRs. This is a very specialized topic, and so we will not cover this type of chart here.

Visual navigation charts, such as the portion shown in Figure 10.3, are designed for navigation under visual flight rule conditions. Although we will describe VFR charts here, we will make no attempt to provide instructions for reading them. Anyone planning to pilot a plane must obtain specialized training. Here we are concerned with what such maps are and what they show. The Resources section at the end of the chapter has some references for those who want more detailed information. VFRs require that the pilot be able to see the ground and control the plane's altitude and be able to see and avoid obstacles. Several series of visual charts are published at different scales to provide accurate and up-to-date information. These series are Sectional Charts (1:500,000); World Aeronautical Charts (1:1,000,000), and Jet Navigation Charts (1:2,000,000). Like symbols for nautical charts, those for aeronautical charts are available in a booklet at *http://aeronav.faa.gov/content/aeronav/online/pdf_files/ VFR_Chart_Symbols.pdf.*

The major features shown on the legend of VFR aeronautical charts are of three types: topographical, obstructional, and aeronautical. Terrain is shown on VFR charts by contour lines, shaded relief, and altitude tints. Many landmarks that can be recognized from the air are identified by a description next to a small symbol, such as a black square for a building, a circle for oil wells, and a black dot for water, oil, or gas tanks. Obstruction symbols are used to represent human-made vertical features,

FIGURE 10.3. Visual navigation chart.

usually those that are more than 200 feet above ground level. Features below this height are shown if they are considered hazardous obstructions, such as antennas, oil or water tanks, lookout towers, and smokestacks. The National Aeronautical Charting Office (NACO) has a file of over 90,000 such obstacles in North America, the Caribbean, and Mexico. The maps also show the maximum elevation figure (MEF), which is the highest elevation including terrain and obstacles within a quadrant. The MEF is especially important on aeronautical charts. A tragic example in recent years was that of a low-altitude military flight in Italy that cut the cables of a ski lift that wasn't shown on the pilot's map, resulting in the deaths of 20 people.

VFR charts also include such aeronautical information as radio aids to navigation and, of course, airports. Airports are divided into public use and military; the map legend designates the type and length of runways and the services available. *The Aeronautical Chart Users Guide* listed in the Further Reading section has a complete listing of symbols.

ROAD MAPS

Road maps are ubiquitous. They are the maps most people are familiar with and probably think of when they see the word *map*. The term *road map* is even used as a metaphor for a plan or program design, such as, a "road map to the future." Until recently, automobile road maps were paper maps, but with the introduction of GPS and online maps, are paper road maps obsolete? Have they been replaced by these newer products and technology? While GPS and online maps may seem a world apart from the humble paper road map, the methods of reading and interpreting them are much the same. Thus, here we first look at conventional road maps, and then we discuss GPS for automobile travel and online products such as GoogleMaps, MapQuest, Bing, and Navtcq, showing how using these differs from paper maps.

In order to get a better comparison of the old and new technology, part of this book was written on a cross-country road trip armed with an in-dash GPS, a set of state road maps, and an up-to-date road atlas. Not surprisingly, each of these tools had strengths and weaknesses in navigating. In planning routes, the road atlas was useful, while on the road, the state maps kept us on course; the GPS gave us an instant view of where we were, and its turn-by-turn voice directions helped us find our motel for the evening. Along the way, the POI (points of interest) feature was invaluable in finding gas stations, restaurants, and motels. There are techniques and skills for using each of these tools.

A Brief History

Road maps have probably been in existence ever since there have been roads. Certainly they predate the automobile. Native Americans, including the Inuit of Alaska and pre-Columbian peoples in Meso America, had maps for way-finding; the Meso American maps showed roads with a series of footprints. There are maps in museums that show pilgrimage routes in the Middle Ages, 18th- and 19th-century carriage route maps in England that show mileage, and even handkerchief maps from the

18th century that show distances from London for travelers in England. In America, Christopher Colles produced *A Survey of the Roads of the United States* in 1789. The automobile road map was preceded by bicycle route guides and maps during the bicycle craze of the 1890s. The earliest way-finding tools for automobiles were itineraries or route guides; there were no numbered roads, so navigation was from point to point with road distances. The Road Map Collectors Association dates the first automobile road map as 1898. Rand McNally was a pioneer with its Auto Trails Series that showed photographs of intersections on the maps. The American Automobile Association began to publish road maps in 1911. By the 1920s, the road map in its now familiar form had developed. From the 1920s to the 1980s in America, road maps were given away at gasoline stations as a form of advertising; the illustrations on the map covers promoted the oil company as well as tourism. Several mapping companies, General Drafting, Gousha, and Rand McNally, created these free maps. The gasoline crisis of the mid-1970s marked the beginning of the end, and by the 1980s, with the exception of official state road maps, they are no longer free. Now road maps are available through state transportation departments, the American Automobile Association, and private companies such as Rand McNally and the National Geographic Society.

What Road Maps Show

As we have seen, maps are usually created with a specific purpose in mind, and this is especially the case with road maps (Figure 10.4). They are designed for a particular user, and anything considered extraneous to the automobile traveler is omitted. Thus, the basic features shown are, of course, roads and their classifications, such as

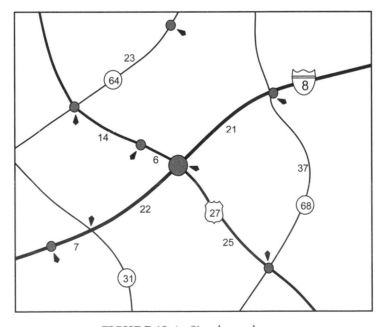

FIGURE 10.4. Simple road map.

freeway, divided road, two-lane, dirt, and the like; highway numbers for interstate, federal, state, and county routes; and distances between points. Cities and towns are shown, often by a symbol or type font that indicates the approximate size. Rivers are usually shown, but railroads may be shown or omitted depending on the publisher. Most road maps include an index of places using an alphanumeric grid.

Other features that might be found on road maps are scenic routes and points of interest, such as historic sites and museums. Elevations of high points may be indicated, but if the nature of the terrain is shown, it is usually with some form of relief shading, not contour lines, because the motorist is usually not interested in detailed elevation information. On AAA maps, towns with AAA-approved lodging are shown in a different color from other towns. Frequently, larger-scale inset maps of cities are included to aid the driver in navigating urban areas. Insets of national parks and state parks are also common. Two other features frequently found are a table showing distances between cities and towns and/or a map showing driving times schematically (Figure 10.5).

Highway Numbering Systems

Most countries now have numbered highway systems. Here the emphasis is on U.S. highways, but we will briefly discuss the systems in Canada, Mexico, Germany, and the United Kingdom. In the early days of automobile travel in the United States, between 1900 and 1917, a number of "trail associations" or "highway associations" designated existing roads as named routes and "auto trails," such as the Lincoln Highway, the National Old Trail, and the Dixie Highway. The Lincoln Highway, probably the best known, went from coast to coast. Towns would pay to have the route go through their business district, with the idea that increased traffic would

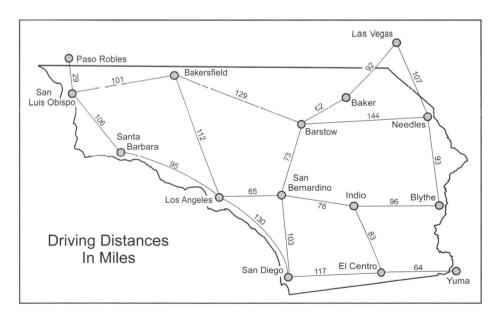

FIGURE 10.5. Driving distance schematic.

bring increased business. Distinctive signs were put on roads that were a part of the route. Some of these roads are now commemorated on current roads that follow the old routes, and there are books and groups of afficionados for the old named roads, such as the Lincoln Highway Association.

Part of reading road maps in the United States is understanding the route numbering systems; why they are numbered as they are; and what the numbers mean. The United States has two national numbered highway systems as well as state and county road networks.

The first highway system in the United States was the Federal Highway System, which was established in 1926 after some years of planning. These are also called the U.S. Highways or U.S. Routes. This system was not "built from scratch"; rather, existing roads were integrated into the numbering system. Within the federal system, north–south routes were given odd numbers beginning with Highway 1 along the east coast and ending with Highway 101 along the west coast. The even-numbered routes went from east to west with U.S. 2 in the North and U.S. 80 in the South. Three-digit numbers are considered spurs of the parent highway except for Highway 101, which is actually considered a two-digit number, in which 10 is the first digit (10-1). Spurs do not necessarily connect with the parent highway. Some of the most famous routes include Route 66, which began in Chicago and ended in Santa Monica, California, at the Pacific Ocean and U.S. 40, which followed the old Lincoln Highway from New York City to San Francisco. Many of the old federal routes, including Route 66, have been decommissioned, with only segments of the original route remaining and often designated as "historic." Federal highways are identified by a black-and-white shield emblem (Figure 10.6).

The interstate system, formally known as the Dwight D. Eisenhower National System of Interstate and Defense Highways, is a system of limited-access highways that traverse the country as multilane roads. It was first conceived of during the Eisenhower administration and was designed to be able to carry large military equipment. There are now nearly 50,000 miles of interstate highways. The numbers are displayed on a red, white, and blue shield (Figure 10.7). The numbering system of interstate highways is the reverse of the federal system. The interstates begin with I-5 running north to south on the west coast and I-95 on the east coast. The east–west routes begin with I-8 in the South and end with I-94 in the North. Route numbers divisible by five, such as I-10, I-75, I-95, are considered major arteries. It must be remembered that I-66 is *not* the same route as the famous U.S. Route 66, and I-40 is not the same as U.S. 40. Alaska, Hawaii, and Puerto Rico also have interstate highways, but they do not follow the same numbering system and their route numbers have the prefixes A, H, and PR, respectively.

FIGURE 10.6. Federal highway shield.

FIGURE 10.7. Interstate highway shield.

In urban areas of the contiguous states, three-digit numbers designate radial, ring, or spur routes to the interstate system. Ring (circumferential or loop) routes are given an even first digit, and the last two digits indicate the main road, such as 210, 405, 495 that join I-10, I-5, and I-95, respectively. Spur roads are roads that do not rejoin the main route and are given an uneven first digit, such as 510, 110, 710, indicating that they are spurs of I-10 (Figures 10.8 and 10.9).

Along the roads are mile markers that allow emergency workers and others to identify points on the road. These begin at the southern border of a state for north–south routes and at the western border for east–west routes. Traveling north through Colorado, for example, I-25 goes from 0 at the southern border with New Mexico to 299 at the northern border with Wyoming; I-70 goes from 0 at the Utah–Colorado border to mile marker 450 at the Kansas state line. The interstates are limited-access routes, meaning there are no physically intersecting crossroads, but a series of entrances and exits that permit access. Access roads are named and/or numbered. Increasingly, a standard method of exit numbering is used that is based on the nearest mile marker. Thus in our Colorado example, the exit closest to the Utah border is at mile marker 2 and is thus called exit 2. The exit closest to Kansas is exit 438. If more than one exit occurs within a given mile, they are labeled with the mile and a letter, such as 25a, 25b, and so on. Exit numbers are marked on road maps.

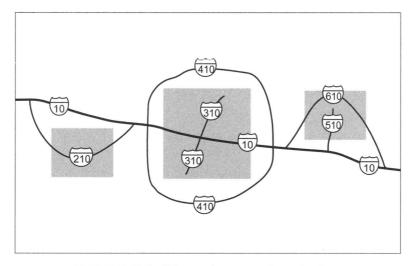

FIGURE 10.8. Ring and spur road numbering.

FIGURE 10.9. Interstate numbering in Los Angeles areas.

Originally, there was no consistency in identifying exits. In California, for example, exits were simply given the name of the street or road they led to. Other states simply numbered the exits sequentially without regard to mile markers, but most states are now converting to the mile marker system and some show both old and new numbers. New York State, however, has not made the conversion, and the numbers are sequential and not based on mileage. For ring routes and spurs, the numbers begin at the start of the route.

In addition to the federal highways and interstates, individual states have numbered state road systems and numbered county roads. In 1918 Wisconsin was the first to develop a numbered state highway system. Each state has a distinctive route sign; some of these signs, notably those in Ohio and Texas, are outlines of the state map (Figure 10.10). Detailed information about each state's highway systems is available online usually through the state's department of transportation.

In Canada the *Trans-Canada Highway* is a system that joins the provinces of Canada from east to west; it is one of the world's longest national highways. Two routes are considered a part of this system: the Trans-Canada, marked by white on green maple leaf route signs, and the Yellowhead Highway which is primarily in the western provinces and was originally marked by a symbol of two pine trees and a yellow man's head. The Yellowhead is to the north of the main Trans-Canada and is designated as Route 16.

German autobahns are a federal limited-access system that began in the 1930s as high-speed roads. The German term is *Bundesautobahn*, or federal motorway.

FIGURE 10.10. Ohio road shield.

There are 12,845 kilometers (7,982 mi) of these roads. The main autobahns that go across the country are single-digit; shorter routes that connect main cities or regions have double-digit numbers based on nine regions. Ring roads and the like are given a three-digit number. North–south autobahns are given odd numbers, and the numbers increase from west to east; east–west autobahns are given even numbers, with lower numbers in the north and higher in the south.

Great Britain has three main categories of roads: M, A, and B. M roads are motorways, which were introduced in the late 1950s. They are the equivalent of interstates and are limited access; the numbering system is based on six zones radiating from London; thus the roads are numbered M1 through M6. There are shorter motorways that, like the U.S. interstate system, are spurs of the main system. A roads are main roads and are also numbered according to zones and also radiate from London; B roads are local roads.

The federal roads of Mexico are divided into high-speed limited-access roads and low-speed open-access roads. The numbering system begins in the northwest, with even-numbered roads east–west and odd numbered roads north–south.

Road Classifications

Most road maps show various classifications of roads. While the United States Department of Transportation has issued formal classifications designating roads as arterials, local roads, and collector roads, most road maps distinguish road types as principal roads, through roads, or secondary roads, local, including streets, and unpaved. Principal roads may also be identified as divided or undivided, and toll roads are usually indicated. A symbol, such as a series of green dots paralleling a road on AAA maps, may indicate scenic routes.

Road Map Reading Tasks

Reader's Purpose

You may have several reasons for using a road map. Do you want to plan a cross-country trip, find a route within a city, calculate distance to the next town, or follow a route to a specific destination? Are you using a city map, state map, or road atlas? The scale and detail shown on these tools differ and therefore, the amount of information will be different.

The simplest task is determining the *distance between points.* Although road maps usually have a printed scale and they can be used to calculate distances, most road maps also show distances between marker points, which are usually intersections or towns. The symbols for the markers are shown in the legend (Figure 10.11). One simply adds up the numbers along the route to determine the total distance. There are also map-measuring tools that consist of a small wheel and a readout. One sets the tool to the scale of the map and rolls the tool along the selected route. A danger in using this method is underestimating distance in hilly areas. We know that maps are generalized, and this is one example. Because of map scale, it may not be possible to show every twist and turn of a mountain road; the road is shown as winding, but it is generalized. Thus, following the road with a measuring device will not give an accurate readout. If one is using a map on the computer, such as Google Maps or Map Quest, the program will calculate the distance and driving time for you, as will an in-car navigation system. However, one should be able to calculate distances on a conventional map since the other options may not be available at a given time.

A more complex task is route planning. While programs such as Google Maps can provide directions between places on opposite sides of the country, the route is normally the most direct and quickest. For a long road trip, some people are not interested in the quickest, but may have some sightseeing plans or want to visit friends or relatives along the way. Some programs do provide such options.

Map Currency

The date of the map is important. Many changes can take place even in a 5-year period: Previously unpaved roads may now be paved; other roads may have been closed; bridges may be closed and new routes added. Road maps usually include the publication date somewhere on the map but often not prominently.

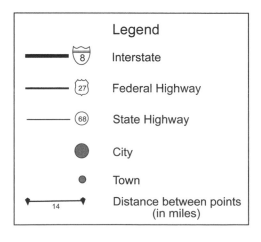

FIGURE 10.11. Simple road map legend.

Caveats

Road maps may include a warning statement that road surfaces might vary or that certain roads are not suitable during different times of the year. Like the publication date, this warning is not always displayed prominently. The user should be aware that in using *any* map there can be errors or changes. When in doubt, it is always advisable to get local information by asking local people, by calling the office of transportation, or by going online. Map publishers frequently provide "important phone numbers" for transportation information on the map.

Orientation

Which way is up? In other words, how do you orient a road map for reading, north at the top or in the direction of travel? This question as regards map reading has been responsible for many arguments, insults, and frustration. There are those who insist that the only proper way to read a map is to keep north at the top and that any other way is wrong. Others, equally insistent, believe that the map should be oriented in the direction of travel and that such orientation leads to fewer errors when turning right or left. As we have seen, north at the top came about after the invention of the magnetic compass. For reading road maps, deciding which way to hold the map is largely a matter of convenience, preference, or how one was taught. Either approach works, and in-car GPS units usually permit either orientation.

Folding Road Maps

For some reason folding road maps causes many people great concern. When using a road map, most readers open it to show the area of interest, and over the course of a trip, the map is folded and refolded in a variety of ways. To return it to its (almost) pristine condition, fold it accordion style, attempting to find the original creases, and then fold roughly in thirds.

Road Atlases

Road atlases are collections of maps of an area or country on a smaller scale than the usual road maps; commonly, for the United States, they show each state on one or two pages, with perhaps an inset of cities or other congested areas. They normally also include a U.S. map that shows major routes and a chart of distances such as the one shown in Figure 10.12. Because of scale limitations, road atlases do not include as much detail as a conventional road map and are best suited to overall route planning. The U.S. map is used to plot the primary route from state to state, and the state maps are used to plot a route through the state and perhaps side trips to scenic spots. The city insets can be useful to find one's way through a tangle of freeways to the correct exit. Basically, road atlases provide a good overview and are convenient to use while planning a trip at home and a quick reference while on the road.

Using Internet Maps

Several companies provide online maps and mobile apps, including AAA Triptik, Google Maps, Google Earth, Map Quest, Bing, and Navteq; all are quite similar in appearance and use. The locational data on these sites is provided by Navteq or Tele Atlas, with the exception of Google, which has its own database. The most basic use, as we have seen, is finding a location that might be an address or a point of interest (POI). POIs are restaurants, motels, gasoline stations, museums, schools, shopping malls, and the like. For navigation, these tools also will provide a route map and turn-by-turn directions; the directions also give estimated time for the trip, which is useful for planning. Some provide two or three different route options, and most will allow the user to specify types of route, such as no freeways and the like. It is possible to plan a long-distance, even cross-country trip, using an Internet map, but though useful, as noted earlier, it is probably not the best tool.

Generally, online maps are read in much the same way as conventional maps, although they do have the advantage of zooming for larger and smaller scale. However, one must pan back and forth to get the whole picture or zoom out to a smaller scale. The map size and scale are limited by the size of the computer screen. Thus, to see an entire city, the scale becomes too small to see detail, unlike a paper map that can show the entire area on a large piece of paper. Online maps usually have options of map, hybrid, and satellite views (although the satellite views are largely low-level aerial photographs) and provide turn-by-turn directions for automobile, walking, and public transit. Google also provides directions for bicycles.

	Albuquerque	Carlsbad	Clovis	Farmington	Gallup	Los Alamos	Roswell	Santa Fe	Silver City	Tucumcari
Albuquerque	0	275	220	178	139	77	199	56	235	173
Carlsbad	275	0	186	451	411	301	75	267	312	257
Clovis	220	186	0	397	357	247	110	213	403	79
Farmington	178	451	397	0	102	153	376	205	389	350
Gallup	139	411	357	102	0	229	336	197	269	311
Los Alamos	77	301	247	153	229	0	227	34	331	201
Roswell	199	75	110	376	336	227	0	192	246	157
Santa Fe	56	267	213	205	197	34	192	0	223	166
Silver City	235	312	403	389	269	331	246	223	0	394
Tucumcari	173	257	79	350	311	201	157	166	394	0

NEW MEXICO ROAD DISTANCES

FIGURE 10.12. Distance chart.

Because these tools are online, many people tend to think they are more up to date and accurate than a paper map. This is not always the case. The same caveats and limitations occur with these products as with any map. While they are generally current, you should pay close attention to the following typical statement included with the directions:

> These directions are for planning purposes only. You may find that construction projects, traffic, weather, or other events may cause conditions to differ from the map results, and you should plan your route accordingly. You must obey all signs or notices regarding your route. (Google Maps)

Any road map, whether conventional or online, is only as accurate as its data.

A useful tool on Google Maps is the street view of an address. In addition to pinpointing the address, a picture of the building found at the address helps the driver to recognize the place.

GPS in Automobiles

GPS units for automobiles are either built into the dashboard or are available as after-purchase units that plug into the auxiliary outlet and can be mounted on the dashboard or windshield. Hand-held units or a GPS app on a smart phone can also be used, although they are not specifically designed for automotive use. In the near future, in-dash units will probably become standard on all cars. While GPS is a wonderful tool as we saw in Chapters 7 and 8, as with paper maps and online maps, caution must be used.

GPS is not a map, but as we learned in Chapter 7, it is a series of satellites that allow the user to locate him/herself. Current GPS units for automobiles have maps on the screen created from Navteq or Tele Atlas. The kinds of information displayed on the screen will vary with the make and model of the device. Some provide traffic information and allow for detours, and most allow change of scale (i.e., zooming in and out), and change in orientation (i.e., north up, or direction of travel up). Being able to zoom in and out on the GPS map is a real advantage when you are in an unfamiliar area. Zooming out provides the big picture, and zooming in shows the detail of streets and intersections.

A common task when using an automobile GPS is finding an address or POI in an unfamiliar area and a route to get there. Most GPS units allow the user to select the type of POI, such as restaurants and symbols that will be shown on the map. The type of restaurant may also be shown, such as American, French, or Mexican. You may also put in a restaurant's name, and the unit will display its location and, if the option is selected, provide a route. For chain restaurants, such as McDonald's, Subway, Denny's, and others, inputting the name will show all of the chain's locations on the map, allowing the user to select the closest, or the one nearest the final destination. So that the driver does not have to constantly look at a list of directions, voice guidance is provided as verbal turn-by-turn directions, such as "in point 5 miles turn left." The directions are also on the screen.

Caveats

Four major items need to be considered before using an automobile GPS. The most important is not being distracted by looking at the unit instead of the road or trying to enter addresses while driving; some in-dash units will not permit the latter while the car is in motion as a safety feature. Second, although few people do, read the manual for your unit. It will help you get the maximum benefit from the device. Third, *update* the database at least yearly. Just as an out-of-date paper map can mislead, so can an out-of-date GPS database. The satellites do not beam a map to your unit. Depending on the unit, updating can be done through the automobile dealer for in-dash units, or by downloading an update for plug-in units. Fourth, heed the notices, warnings, and other information usually highlighted or boxed in the instruction manual; these tell you the limitations or problems you might encounter. One such warning is: "For some areas, the roads have not been completely digitized in our database. For this reason, the route guidance may select a road that should not be traveled on" (Toyota user's manual, 2011).

It is the last caveat that has led to the greatest problems with automobile GPS units. Databases are not flawless; they may show closed roads as passable, they may show dirt roads as paved, or they may show proposed roads as already existing. A particularly tragic example was featured in an article in the *Sacramento Bee*, January 30, 2011, titled "Death by GPS." In this story, in Death Valley in August of 2010, a woman and her child followed a road shown on their in-car GPS. Unfortunately, the road was not passable; they became stranded without water and the child died. Sadly, this is not an uncommon occurrence and does not occur just in deserts; it can also happen on mountain roads closed by snow or other roads that are open only during certain times of the year. People who might question a paper map seemingly place blind faith in the technology of GPS. It is a useful tool, but like any tool, it should be used wisely and appropriately.

BICYCLE MAPS

As we noted earlier, bicycle maps predate automobile maps, and bicycle routes were in existence in the 1890s. The coming of the automobile largely eliminated bicycles as transportation except in very poor areas and for children. However, bicycle transportation has experienced a resurgence, brought about in part by "Bikecentennial" in 1976 when many cyclists traversed the United States in celebration of the nation's bicentennial. The oil crisis of the 1970s, downturns in the economy, and environmental concerns led to more interest in bicycle commuting. As a result, cities and states are designing and promoting bicycle routes.

Several different agencies and companies are involved in creating modern bicycle maps for the commuter and bicycle tourist. In addition, there are user-created routes on websites, such as *mapmyride.com*. State departments of transportation are responsible for creating routes and maps; larger and midsize cities, such as New York, Chicago, Portland, Oregon, and Long Beach, California, have created maps of bicycle routes as both printed maps and online. The majority of these maps are aimed at the bicycle commuter and recreational cyclist. For cyclists interested in long-distance touring, such as trips across the country, Adventure Cycling Association, a nonprofit organization, has created a series of 22 routes covering over 42,000 miles with accompanying maps. The long routes are covered by multiple maps. For example, the Southern Tier from San Diego, California, to St. Augustine, Florida, requires seven maps to cover the 3,058-mile route, and the 4,221-mile Trans America route requires 12 maps (Plate 10.1). The American Association of State Highway Transportation Officials (AAHTSO), in conjunction with Adventure Cycling, has proposed a bicycle highway network akin to the national highway system. Several companies make cycling maps of small areas. Google Maps has introduced cycling routes as an option on their online maps.

As with other maps, cycling maps are designed for a specific purpose, and the maps show features of interest to cyclists (Figure 10.13). Elevation and gradient are of particular interest. Therefore, such maps either indicate the percent gradient with color or provide a series of profiles for each segment (Figure 10.14). Distances are of greater consequence on a bicycle than when driving. Ten miles is of little consequence in an automobile, but on a bicycle it can mean a half hour or more of pedaling, depending on the terrain, so distances are provided more frequently. Symbols indicate POIs for the touring cyclist, such as bicycle shops, hotels, hostels, campgrounds,

FIGURE 10.13. Percent gradient is important for cyclists.

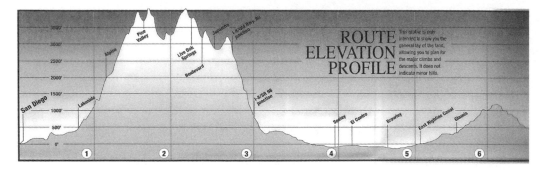

FIGURE 10.14. Profile for cycling. *Source.* Adventure Cycling Association. Reprinted by permission.

groceries, libraries, and other services. On Adventure Cycling maps, the orientation is toward the direction of travel, not magnetic or true north, in order to fit a greater area on the restricted size. The maps are designed to fit on a handlebar case.

GPS units especially designed for bicycles are available, and, of course, handheld GPS units can be used. The simplest bicycle GPS units give location, speed, and distance traveled, but they do not have a map; more expensive units have built-in maps. The same caveats apply to these units as to units for automobiles.

FURTHER READING

Akerman, James R. (Ed.). (2006). *Cartographies of Travel and Navigation*. Chicago: University of Chicago Press.

Calder, Nigel. (2003). *How to Read a Nautical Chart*. Camden, ME: International Marine/McGraw-Hill.

Federal Aviation Administration. (2012). *The Aeronautical Chart Users Guide* (9th ed.). New York: Skyhorse.

Huth, John Edward. (2013). *The Lost Art of Finding Our Way*. Cambridge, MA: Belknap Press of Harvard University.

Lewis, Tom. (1997). *Divided Highways*. New York: Penguin Books.

Taylor, E. G. R. (1971). *The Haven-Finding Art: A History of Navigation from Odysseus to Captain Cook* (augmented ed.). New York: American Elsevier.

RESOURCES

AAA history
    www.aaa.com/aaa/074/centennial/webpages/maphistory.html
Bicycle Maps
    adventurecycling.com
Nautical Charts
    www.nauticalcharts.noaa.gov/mcd/chartno1.htm
    www.celebrating200years.noaa.gov/foundations/nautical_charts

Road Maps Collectors Association
roadmaps.org

Internet map sites

AAA (available to members)
www.aaa.com

Bing
www.bing.com/maps
Google
maps.google.com
Map Quest
www.mapquest.com
Navteq
www.navteq.com

CHAPTER 11

Maps for Special Purposes

The importance of flat maps is that they can be
drafted to suit special needs.

—*Mapping* (p. 113)

The types of maps in this chapter are widely used but are aimed primarily at a specific
type of user, the geologist, meteorologist, astronomer, tourist, or for a specific effect,
such as persuasive maps. Maps for navigation, whether sea, land, or air, can also
be considered special purpose, but because they form such a large group they were
treated separately in this book. While geological, astronomical, and weather maps
require some specialized knowledge for most effective use, the general map user can
gain a great deal of information from them.

GEOLOGIC MAPS

Geologic maps are designed to show different kinds of rocks, geologic features, and
earthquake faults; the geologic features are superimposed over a base map. As with
most maps, the features are symbolized with colors, lines, and dots. Geologic maps
are comparatively new on the cartographic scene, dating from the beginning of the
19th century.

Geologic map reading, like aeronautical and nautical chart reading, is a specific
type of map reading, and to get the maximum amount of information from these
maps, some background in geology is desirable. However, the average map reader,
using basic map reading skills, can learn much from these maps. The uses of geologic

maps depend on the scale, but include location of economically important mineral deposits, estimation of resources, and planning of engineering projects (Plate 11.1).

Geologic maps in the United States are produced primarily by the U.S. Geological Survey (USGS), which is also the major topographic mapping agency. Most states also have geological surveys and produce maps. The scales of geologic maps are similar to those of topographic maps, with 1:24,000 and 1:100,000 being large scale, but small-scale maps may have scales of 1:750,000 or smaller.

WEATHER MAPS

One of the most commonly seen kinds of map is the daily weather map. Most newspapers include a version of this map, and news broadcasts feature a local weather segment illustrated with weather, satellite, and radar maps; The Weather Channel is familiar to most people and is a popular channel with travelers. We must distinguish between climate maps and weather maps. *Climate* is the temperature, rainfall, wind, and pressure of a place over the long term, and climatologists have established climate types that may be shown on maps. *Weather* refers to the day-to-day or present conditions of a place. Thus, climate maps distinguish between humid climate regions, arid regions, and the like, and maps that show these climates or maps of average rainfall or average temperatures are grouped with thematic maps. In the United States, weather maps are prepared on a daily basis by the National Weather Service division of the National Oceanic and Atmospheric Administration (NOAA) (Figure 11.1). Like geologic maps, to get the maximum information from weather maps, some background knowledge is needed—in this case, some knowledge of meteorology. In general, the daily weather map shows barometric pressure with isobars, which are a type of isarithmic line: weather fronts, which are boundaries between two air masses, with linear symbols as shown in Figure 11.2; and wind direction with arrows. Weather maps of the United States in newspapers show symbols for kinds of precipitation and use colors based on isotherms to represent temperatures. Television weather maps, now frequently in "real time," show radar images of precipitation for the area.

TIME ZONE MAPS

Although the Harrison chronometer solved the problem of determining longitude from a prime meridian as we saw in Chapter 4, multiple prime meridians still created confusion. As we have seen from the Harrison story, time is tied to longitude. For centuries, people recognized noon as the time when the sun was at its highest point in the sky, which meant that the sun was directly on the local meridian. In fact, the terms a.m. and p.m. (ante meridian and post meridian) reflect this as before the meridian and after the meridian. This works well for small areas, such as a town, but with rapid steamship travel and the introduction of rail travel, a multitude of different "noons" made scheduling impossible. The International Meridian Conference in Washington, DC, in 1884 adopted the meridian that ran through the Royal Observatory in Greenwich, England, as *the* prime meridian for most countries of the

A

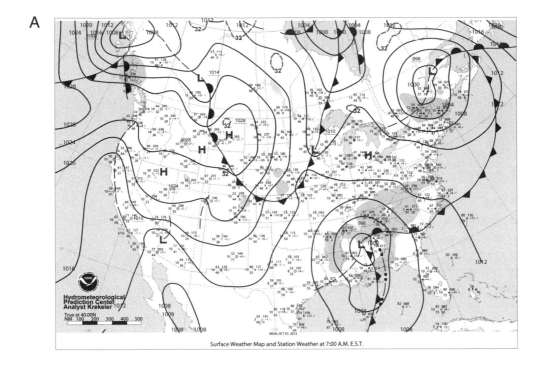

Surface Weather Map and Station Weather at 7:00 A.M. E.S.T.

B

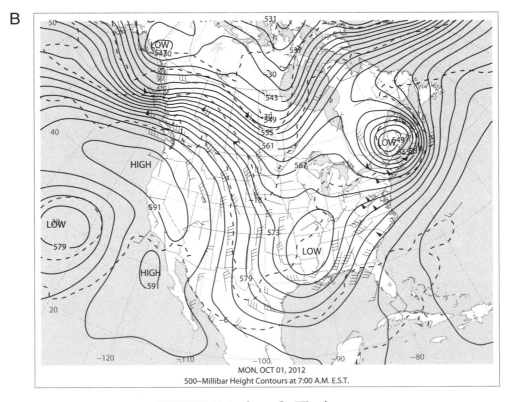

MON, OCT 01, 2012
500–Millibar Height Contours at 7:00 A.M. E.S.T.

FIGURE 11.1a, b, c, d. Weather maps.

C

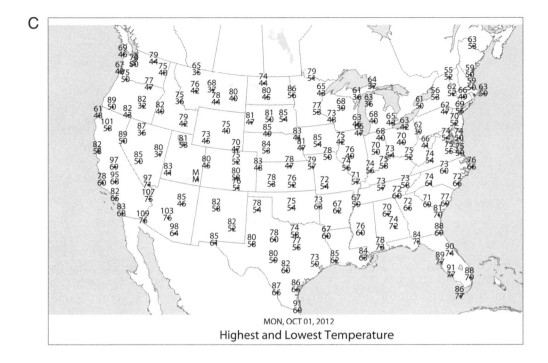

MON, OCT 01, 2012

Highest and Lowest Temperature

D

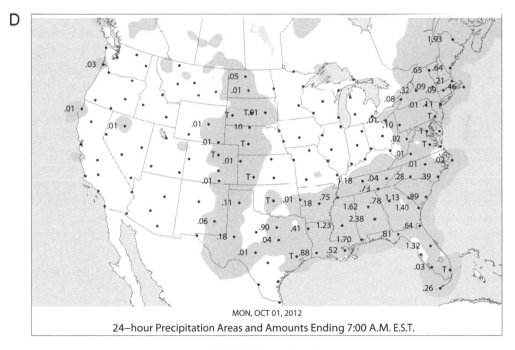

MON, OCT 01, 2012

24–hour Precipitation Areas and Amounts Ending 7:00 A.M. E.S.T.

FIGURE 11.1. *(cont.)*

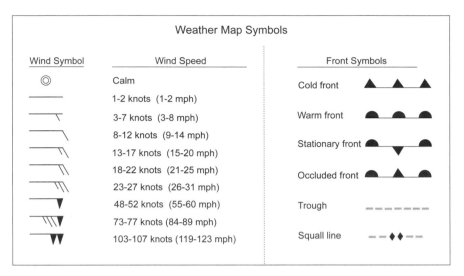

FIGURE 11.2. Weather map symbols.

world, and time zones were introduced in which noon was calculated for a central meridian. France abstained from the vote and continued to use the Paris meridian for some decades. At Greenwich Observatory a brass line (now stainless steel) embedded in the pavement marked the prime meridian and is a popular photo spot for tourists who want to be photographed straddling the eastern and western hemispheres. At night a laser beam aimed northward marks the meridian. A radio time signal is also sent. With air travel a common occurrence, understanding time zones is a necessity.

There is also a line that roughly follows the 180th meridian and is labeled the *International Date Line.* Because the earth is a sphere and rotates on its axis once every 24 hours, there must be a starting line for the day. That line is the International Date Line. The advantage of this meridian is that it passes largely through empty ocean; but the date line doesn't follow the meridian exactly. It zig zags around major island groups and land masses, so that places do not have a permanent one-day difference. If one flies from the United States westward toward Asia on Sunday, when the Date Line is crossed, the day jumps to Monday. Software programs and mobile apps now available show a world map and the time at any place, making it easy to determine if calling your friend in England at 10 P.M. New York time will wake her up.

The International Meridian Conference divided the earth into 24 zones, each 15° wide, but if one looks at a time zone map, it can be seen that the zones usually, except in the oceans, are not exactly 15° wide (Figure 11.3). There are irregularities so that islands in a group are not split or that the eastern half of a city doesn't have a different time than the western half. Some nations do not fit the pattern because they adopted their own times. The time used in each nation is its *legal* or *standard* time. The continental United States has four time zones (Eastern, Central, Mountain, and Pacific); Alaska and Hawaii each have their own zone. We can see from Figure 11.4 that the U.S. time zones also do not follow meridians exactly. This is done in order to keep cities from being split into two time zones.

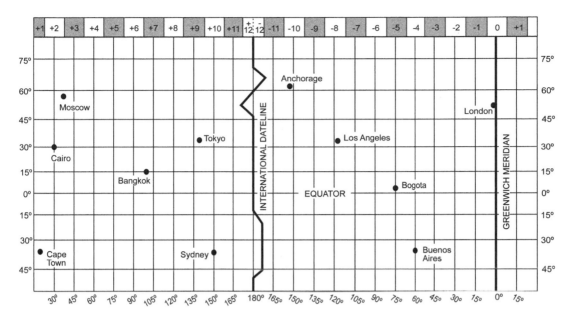

FIGURE 11.3. Schematic map of the International Date Line and time zones.

Some countries change their legal time for part of the year for economic reasons; the change is usually during the summer, and the time is advanced by one hour to take advantage of daylight hours early in the day. The concept was proposed by Benjamin Franklin in 1784 but first implemented during World War I. Most of the United States observes *Daylight Saving Time*. In the United States, Daylight Saving Time extends from the second Sunday in March to the first Sunday in November,

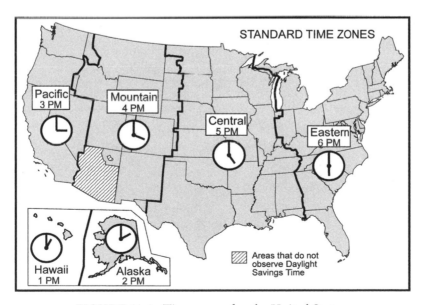

FIGURE 11.4. Time zones for the United States.

which goes far beyond the summer months. Arizona does not observe Daylight Saving Time, and there have been many arguments pro and con regarding the practice.

MAPS OF THE PLANETS AND STARS

Early definitions of the word "map" focused on the earth, and no mention was made of maps of the heavens or other astronomical bodies; yet such maps have been made from quite early times. Arab cartographers made celestial maps as early as the eighth century, the first "map" of the moon was made by Galileo in 1610 (Figure 11.5),

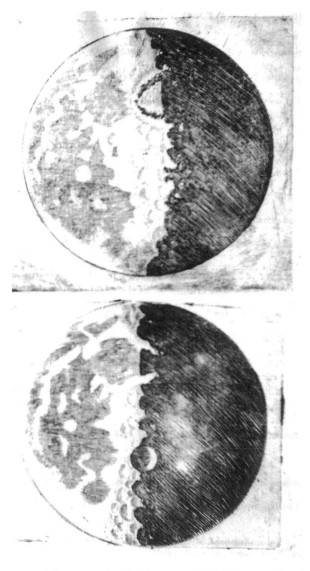

FIGURE 11.5. "Map" of the moon by Galileo, 1610. This is considered the earliest map of the moon (author photograph).

and maps of Mars were made in the 19th century. Galileo's map, which was actually more of a scale drawing, was made one year after the invention of the telescope using a 30-power telescope, less than the common power of many field binoculars today. Maps of Mars were full of speculation in the early days, showing supposed canals and other indications of intelligent life (Figure 11.6).

The impact of space travel has been enormous. Manned and unmanned flights to the moon and unmanned flights around and to Mars sent back, and are still sending back, a wealth of information. Now even elevation and geologic maps of both the moon and Mars are available (Plate 11.2). We noted earlier that Google Moon and Google Mars provide the layperson with imagery and maps of both bodies. The NASA website listed in the Resources at the end of the chapter also has imagery.

Maps of the sky and stars were originally designed for astrology, but commercial star maps are available now that help the amateur astronomer learn the "geography" of the sky: the locations of planets, constellations, galaxies, and stars. Maps are now made from imagery taken from the Hubble Telescope, and these are available on Google Sky. Other computer software and apps are available, such as Starry Night and Star Walk.

HISTORIC AND HISTORICAL MAPS

Historians use maps as research documents. Old maps show spatial relationships in the past, political divisions in a particular time frame, and beliefs about the world. Cadastral maps can show who owned a parcel of land, census maps show where a person lived, and road maps show travel patterns. These maps are all examples of *historic maps*, maps that were made at a particular time and place. These maps are

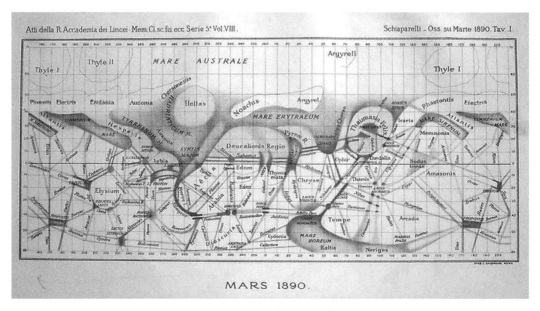

FIGURE 11.6. Early map of Mars.

studied by historians, historians of cartography, and genealogists. Historians, historical geographers, archaeologists, and the like make *historical maps*, which are modern maps made to show historical features and relationships. Although many reproductions of historic maps have been published in book form, there are also historical atlases that re-create an area or event in the past, such as the American Civil War (Figure 11.7).

For the map user who wants to work with old maps, a number of websites are available. David Rumsey's site, *www.davidrumsey.com,* is a searchable site with over 34,000 historic maps of a variety of areas and time periods viewable online (Figure 11.8). Many museums have collections viewable online. The USGS has made historic topographic maps available online; these are quite valuable when doing a historical study of an area or looking at changes in an area. The U.S. Census has recently made the detailed 1940 census available to the public, including property maps, at *http://1940census.archives.gov.* Of course, these maps are read and interpreted in the same way as modern maps.

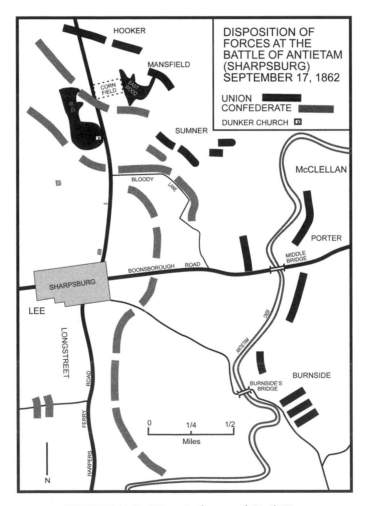

FIGURE 11.7. Historical map of Civil War.

RECREATION AND TOURISM

National Parks and Monuments

Although topographic maps cover most outdoor recreation areas, such as national parks and monuments, national recreation areas, and state parks, it may take as many as 12 or more topographic quadrangles to cover a large national park. While hikers want the detail of such maps, motorists and casual visitors prefer a map of the area on one sheet that will show points of interest, campgrounds, services, and the like. For this reason, all U.S. national parks and monuments are covered by visitor's maps that can be obtained at the entrance of the park or online by going to *www.nps.gov*. Each national park has its own website with map. These can be found at the national parks website: *www.nps.gov*. This site provides the user with a list of national parks and their individual websites, and each park site has maps (Figure 11.9). These maps use a uniform format for all of the parks and monuments, both on paper and online. The online maps permit panning and zooming.

When discussing these recreational areas, it is important to recognize the distinctions between the various kinds of areas. The mission and consequently the uses are different. Yellowstone National Park, established in 1872, is the oldest such park in the world, and since then more than 1,200 national parks have been established in over 100 nations. In the United States, the National Park Service, which is a part of the U.S. Department of the Interior, administers some 20 kinds of parks, monuments, and battlefields, as shown in Table 11.1. As mentioned earlier, national parks are created through acts of Congress, whereas national monuments can be created on federal lands by a proclamation of the president and may later be designated as parks by Congress. There are 401 parks, monuments, and other sites in all of the states

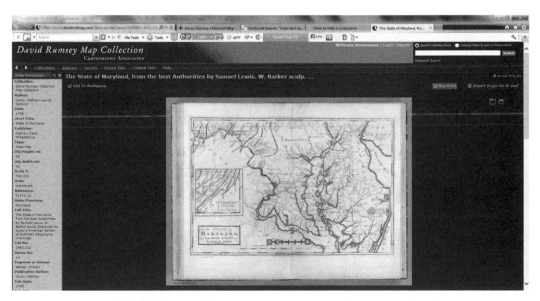

FIGURE 11.8. Example of a map from David Rumsey's online map collection. From *www. davidrumsey.com*. Reprinted by permission.

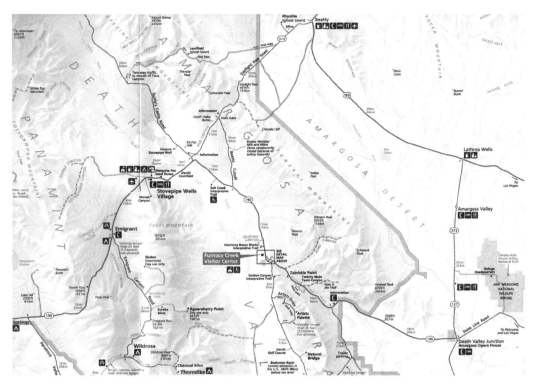

FIGURE 11.9. National park map of Death Valley.

and territories. These sites are protected from logging, mining, hunting, grazing, and other such activities.

National forests, though still federal lands, have a different mission and are administered by the National Forest Service under the aegis of the U.S. Department of Agriculture. Unlike national parks, the national forests permit and even encourage commercial use such as logging and grazing, as well as hiking and camping.

While topographic maps and governing agency maps are the most common maps for these areas, maps for national parks are also commercially available. The National Geographic Society has both paper and digital maps of selected national parks; these are produced in partnership with the National Park Service and other agencies. The digital maps are available on DVD and provide elevation profiles, and three-dimensional flythroughs, and can send waypoints to GPS. The user can also annotate the maps. Other private companies provide specialized maps for various recreational areas for hikers, mountain bikers, skiers, and campers.

Cave Maps

Cave maps are used by scientists studying caves as well as by recreational users called cavers or spelunkers. Speleology is the scientific study of caves, and in order to engage in such study, the first step is mapping the cave. Because of the nature of caves, this mapping presents a special problem. First, caves are three dimensional with passages

TABLE 11.1. U.S. National Park System

Type of designation	Total designations
National battlefields	11
National battlefield parks	4
National battlefield sites	1
National military parks	9
National historical parks	46
National historic sites	78
National lakeshores	4
National memorials	29
National monuments	78
National parks	59
National parkways	4
National preserves	18
National reserves	2
National recreation areas	18
National rivers	5
National wild and scenic rivers and riverways	10
National scenic trails	3
National seashores	10
Other designations	11
Total units	401

Note. Data from National Park Service.

that twist and tilt and "rooms" that are at different levels. Traditionally, however, caves have been mapped in two dimensions. Second, GPS does not work underground, so a tool normally used for mapping the earth's surface and reading maps of the surface is not available. Even magnetic compasses can present problems owing to the nature of the surrounding rocks. Computer software is now available for creating maps after the surveying is done.

The National Speleological Survey organization has a subgroup called the Surveying and Cartography Section and publishes a journal, *Compass and Tape,* that is available online (see Resources at the end of the chapter). This journal includes examples of cave maps. There is an effort to standardize the symbols used in cave mapping, which will make it easier for those using cave maps. An international organization, the Union Internationale de Spéléologie (UIS) has produced a set of standardized symbols (*www.carto.net/neumann/caving/cave-symbols*). Other national organizations, such as the Australian Speleological Society, are adapting these symbols, but some local organizations still use their own symbols (Figure 11.10). Cave maps include several symbol categories, such as survey stations, entrance types, changes of slope or level, height of "ceiling," depth of water, materials and deposits, and artificial constructions.

It is not advisable for a beginning spelunker to attempt following a cave map without an experienced guide.

Maps of Tourist Attractions

When we go to an amusement park, zoo, county fair, or other tourist attraction, we are usually provided a map at the entrance; many facilities now have online maps on

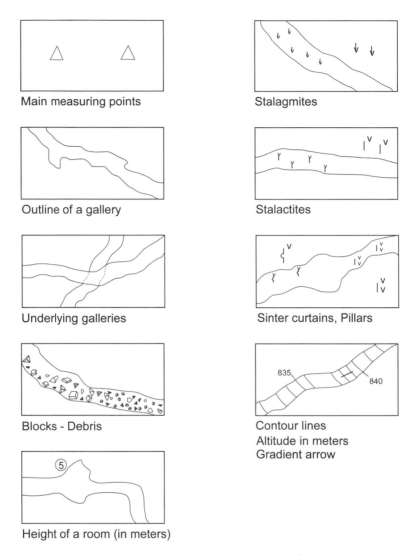

FIGURE 11.10. Cave map symbols.

their website so that you can plan your trip. While the maps vary somewhat with the attraction, there are similarities; maps, as we have learned, are designed for specific purposes and users. Tourist attraction maps are not designed for serious map analysis, but rather to navigate one's way on foot through an often confusing, noisy network of paths in an area that may range from a few acres to more than a square mile. In addition, a child may be the designated navigator. Thus, most such maps are pictorial and in color, with the thrill rides and exhibits shown in small pictures in 3/4 view. While some may include a separate page listing attractions, almost all have a key on the map itself showing the services that the visitor might need or want, such as restrooms, food, first aid, ATM, gift shops, and general information. Many also include wheelchair and baby stroller rental sites (Plate 11.3). A well-designed attraction map allows the visitor to get the maximum benefit from the trip.

MAPS THAT PERSUADE

As we noted in Chapter 1, many maps are designed to persuade the reader to believe something, to buy something, or to do something. *Persuasive maps* are intended to change, or in some way influence, the reader's opinion or beliefs. Persuasion can be viewed as a continuum from minimal (the goal of most topographic maps) to propagandistic. The term *propaganda* has come to mean persuasion with evil intent; propaganda maps are persuasive, but not all persuasive maps are propagandistic. Included in persuasive maps are maps in advertising, maps by special interest groups to "sell" their ideas and philosophies, and maps by government agencies to promote policies. It must be emphasized that just because a map is persuasive it does not follow that the intent is evil and untruth.

In order to persuade, the cartographer uses a number of techniques—feature selection, distortion, symbolization, text, color, and map projection. All of these are legitimate techniques in mapmaking, but here they are used to emphasize the cartographer's intent. For example, selecting only features favorable to the message; using suggestive symbols, such as bomb bursts, and arrows, vicious animals, and nooses; using persuasive or inflammatory text; using unpleasant colors for the "bad guys"; and using projections that distort the area in question are all methods that are used. Often persuasion is signaled by persuasive text. Figure 11.11 is an example of a persuasive map.

The map reader must be aware that maps have agendas. In reading and interpreting any map, the user must exercise critical thinking skills. Maps are powerful tools. In the blunt words of David Greenhood, "People are plain suckers if they never ques-

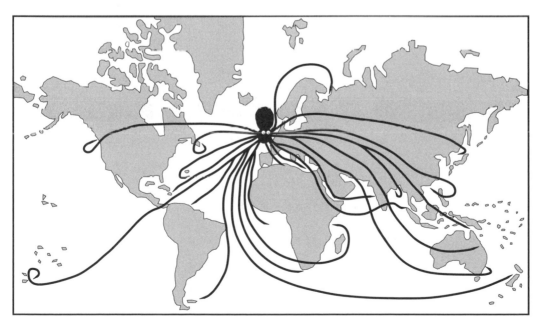

FIGURE 11.11. The octopus has been a popular symbol on persuasive maps for 150 years. This is a re-drawing of a Nazi propaganda map from 1942.

tion the reliability of maps shoved under their noses. The result might be the total loss of civil liberties" (*Down to Earth*, 1944, p. 5).

MAPS IN LITERATURE

While maps in literature do not form a large part of the world of maps and have a different purpose from most maps, they are frequently encountered and have been used for several centuries. Robert Louis Stevenson based *Treasure Island* on a map he had drawn of a fictitious island, and such a map was included with the book. Jonathan Swift's *Gulliver's Travels* includes maps that illustrate the various lands that Gulliver visited that were drawn by one of the most famous cartographers of the day, Herman Moll. Moll also made a map for *Robinson Crusoe*. Some well-known children's books, such as *Winnie the Pooh* and *The Wind in the Willows*, featured maps by a well-known artist of the day, Ernest R. Sheppard. The works of Tolkien have an entire atlas devoted to Middle-Earth. Unfortunately, when these books are printed in inexpensive or paperback editions, the maps may not be included. Maps continue to be used in works of fiction to the present day.

Maps in literature may be of real or fictitious places. Some authors set their stories in real places, and others create their own towns, countries, or, in science fiction, their own worlds. In some cases, the maps are designed to be used, to allow the reader to follow the plot, and may even be mentioned in the story, but in others they are primarily decorative and found on the endpapers of the book. The appearance of these maps varies greatly, from highly pictorial to conventional. The conventional maps usually include a scale and a north arrow, and if they represent a real place, they could be used to navigate the area (Figure 11.12).

Certain genres of fiction tend to have more maps; mysteries and thrillers, historical novels, and fantasy and science fiction works are probably the most common map-illustrated genres, and the popularity of such maps waxes and wanes with publishing fashions.

FURTHER READING

Aberley, Doug. (Ed.). (1993). *Boundaries of Home: Mapping for Local Empowerment*. Philadelphia: New Society.

Fonstad, Karen Wynn. (2001). *The Atlas of Middle Earth* (rev. ed.). New York: Mariner Books.

Kanas, Nick. (2007). *Star Maps: History, Artistry, and Cartography*. Chichester, UK: Springer Praxis.

Monmonier, Mark. (1996). *How to Lie with Maps* (2nd ed.). Chicago: University of Chicago Press.

Monmonier, Mark. (1999). *Air Apparent: How Meteorologists Learned to Map, Predict, and Dramatize Weather*. Chicago: University of Chicago Press.

Monmonier, Mark, and George Schnell. (1988). *Map Appreciation*. Englewood Cliffs, NJ: Prentice-Hall.

FIGURE 11.12. Map: "The Sound of Broken Glass," illustrated by Laura Maestro from *The Sound of Broken Glass* by Deborah Crombie. Copyright © 2013 by Deborah Crombie. Reprinted by permission of HarperCollins Publishers.

Morton, Oliver. (2002). *Mapping Mars: Science, Imagination, and the Birth of a World*. New York: Picador USA.

Post, J. B. (1979). *An Atlas of Fantasy* (rev. ed.). New York: Ballantine Books.

Tyner, Judith. (1969). Early Lunar Cartography. *Surveying and Mapping, 29*(4), 583–596.

Whittaker, Ewen A. (1999). *Mapping and Naming the Moon*. Cambridge, UK: Cambridge University Press.

Winchester, Simon. (2001). *The Map That Changed the World: William Smith and the Birth of Modern Geology*. New York: HarperCollins.

Wood, Denis. (1992). *The power of maps*. New York: Guilford Press.

Wood, Denis. (2010). *Rethinking the Power of Maps*. New York: Guilford Press.

RESOURCES

1940 Census Maps
  http://1940census.archives.gov
Cave Maps: Survey and Cartography Section of the National Speleological Survey
  www.caves.org/section/sacs/SACS/SACS_Home.html

Geologic Maps
www.nature.nps.gov/geology/usgsnps/gmap/gmap1.html
Google Mars Maps
www.google.com/mars
Google Moon Maps
www.google.com/moon
Google Sky Maps
www.google.com/sky
National Geographic Recreation Maps
www.natgeomaps.com
National Geologic Map Data Base
http://ngmdb.usgs.gov/ngmdb_home.html
National Park Service
www.nps.gov
National Parks Mapping History
http://memory.loc.gov/ammem/gmdhtml/nphtml/nphome.html
Weather Maps
www.weather.gov
Time Zone Maps
www.worldtimezone.com

PLATE 2.1. World map, Mesopotamia. © The Trustees of the British Museum.

PLATE 2.2. Hereford *Mappa Mundi*. © The Mappa Mundi Trust and Dean and Chapter of Hereford Cathedral.

PLATE 2.3. Portolan chart by Albini de Canepa, 1489. From the James Ford Bell Library, University of Minnesota, Minneapolis, Minnesota. Reprinted by permission.

PLATE 3.1. Digital elevation model of Mount St. Helens created from Lidar data, USGS.

PLATE 6.1. Salton Sea photographed from *Gemini 5*, August 1965 (NASA photograph).

PLATE 6.2. Earth from Apollo 17, July 1972 (NASA photograph).

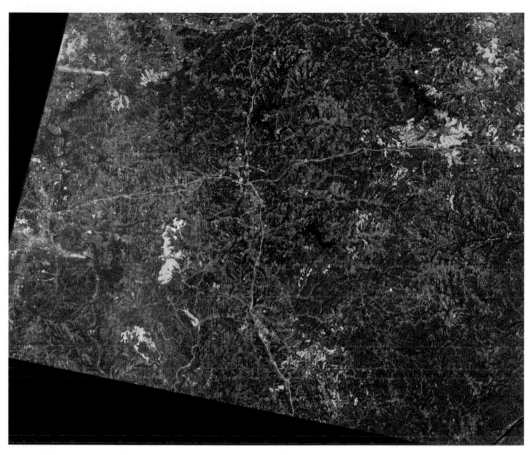

PLATE 6.3. On this false color image of the Muskingum area of Ohio, vegetation appears red and strip mines are shown in blue (USGS photo, Landsat 1, 1973).

Primary highway, hard surface		Boundary: national	
Secondary highway, hard surface		State	
Light-duty road, hard or improved surface		county, parish, municipio	
Unimproved road		civil township, precinct, town, barrio	
Trail		incorporated city, village, town, hamlet	
Railroad: single track		reservation, national or state	
Railroad: multiple track		small park, cemetery, airport, etc.	
Bridge		land grant	
Drawbridge		Township or range line, U.S. land survey	
Tunnel		Section line, U.S. land survey	
Footbridge		Township line, not U.S. land survey	
Overpass—Underpass		Section line, not U.S. land survey	
Power transmission line with located tower		Fence line or field line	
Landmark line (labeled as to type)	TELEPHONE	Section corner: found—indicated	+ +
		Boundary monument: land grant—other	▫ ▫

Dam with lock			
Canal with lock		Index contour	Intermediate contour
Large dam		Supplementary cont.	Depression contours
Small dam: masonry — earth		Cut — Fill	Levee
Buildings (dwelling, place of employment, etc.)		Mine dump	Large wash
School—Church—Cemeteries	Cem	Dune area	Tailings pond
Buildings (barn, warehouse, etc.)		Sand area	Distorted surface
Tanks; oil, water, etc. (labeled only if water)	Water Tank	Tailings	Gravel beach
Wells other than water (labeled as to type)	o Oil o Gas		
U.S. mineral or location monument — Prospect	▲ x	Glacier	Intermittent streams
Quarry — Gravel pit	✕	Perennial streams	Aqueduct tunnel
Mine shaft—Tunnel or cave entrance		Water well—Spring	Falls
Campsite — Picnic area		Rapids	Intermittent lake
Located or landmark object—Windmill	⊙	Channel	Small wash
Exposed wreck		Sounding—Depth curve 10	Marsh (swamp)
Rock or coral reef		Dry lake bed	Land subject to controlled inundation
Foreshore flat			
Rock: bare or awash	*	Woodland	Mangrove
		Submerged marsh	Scrub
Horizontal control station	△	Orchard	Wooded marsh
Vertical control station	BM ✕671 ✕672	Vineyard	Bldg. omission area
Road fork — Section corner with elevation	429 +58		
Checked spot elevation	✕ 5970		
Unchecked spot elevation	✕ 5970		

PLATE 8.1. Topographic symbol sheet. *Source.* USGS.

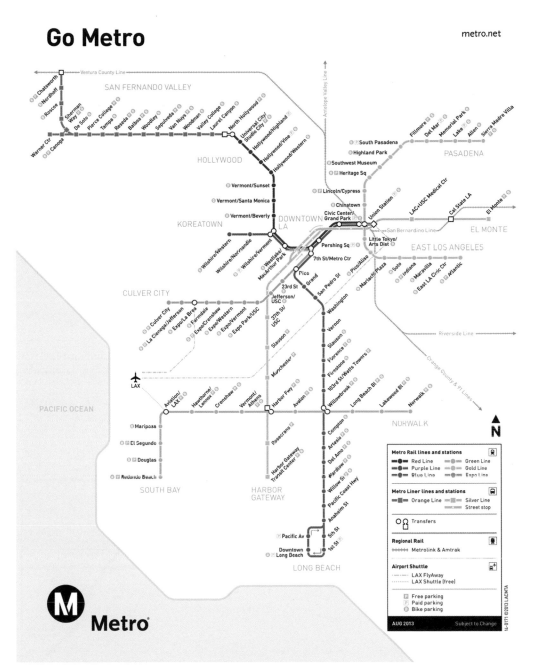

PLATE 9.1. Transportation diagram for the Los Angeles Metro. Map courtesy of Metro (Los Angeles County Metropolitan Transportation Authority).

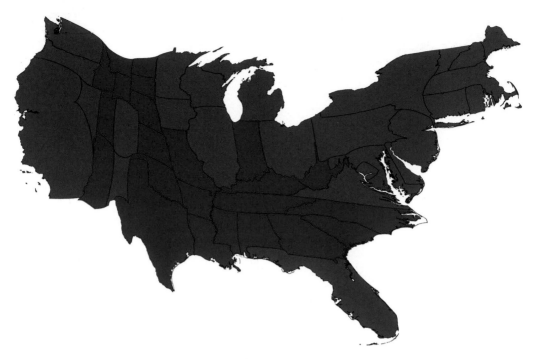

PLATE 9.2. Topologically correct cartogram of the 2012 presidential election results. The size of the state is based on the number of electoral votes. Courtesy of Mark Newman, University of Michigan.

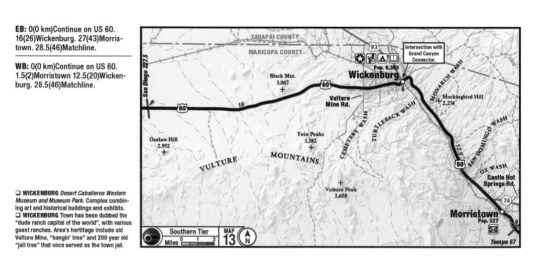

PLATE 10.1. Bicycle map. *Source.* Adventure Cycling Association. Reprinted by permission.

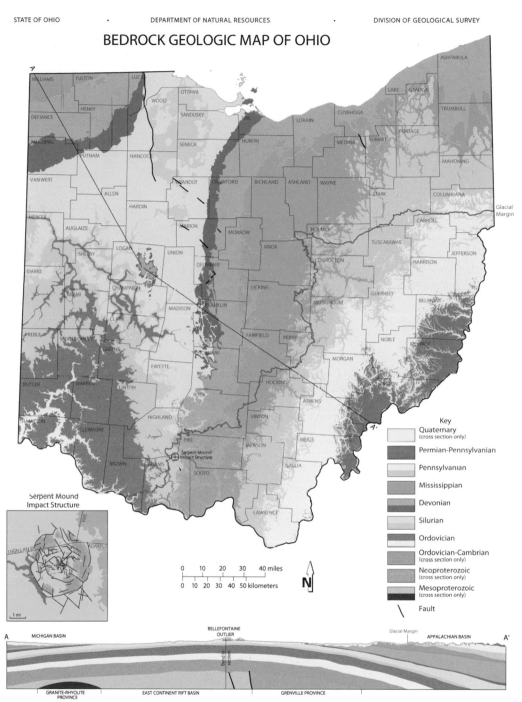

PLATE 11.1. Geologic map.

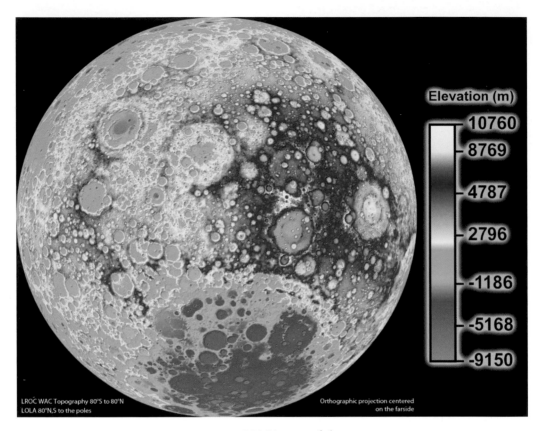

PLATE 11.2. NASA map of the moon.

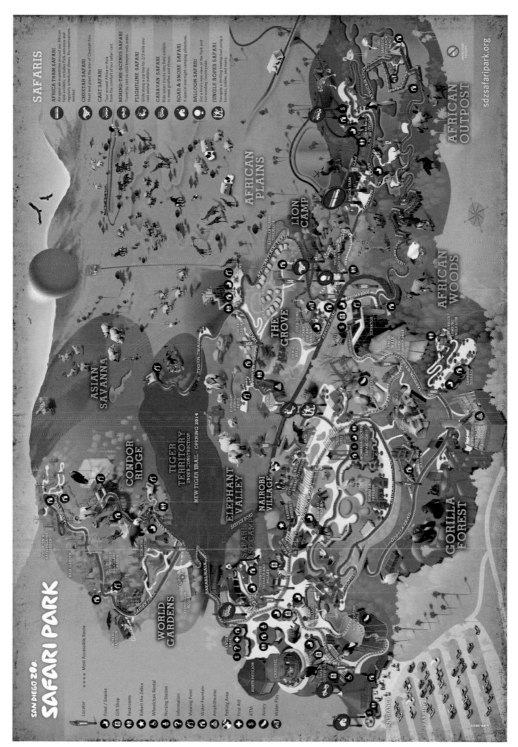

PLATE 11.3. Tourist attraction maps are colorful and pictorial. Map reprinted courtesy of the Zoological Society of San Diego.

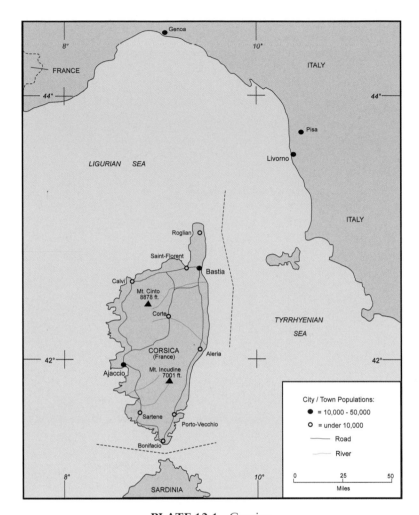

PLATE 12.1. Corsica.

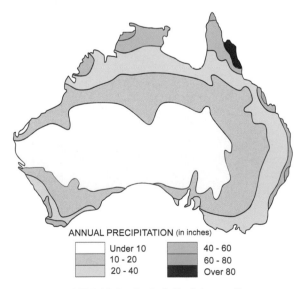

ANNUAL PRECIPITATION (in inches)

Under 10	40 - 60
10 - 20	60 - 80
20 - 40	Over 80

PLATE 12.2. Rainfall of Australia.

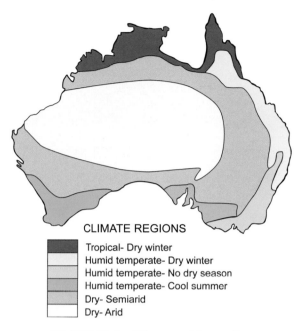

CLIMATE REGIONS

Tropical- Dry winter
Humid temperate- Dry winter
Humid temperate- No dry season
Humid temperate- Cool summer
Dry- Semiarid
Dry- Arid

PLATE 12.3. Climate of Australia.

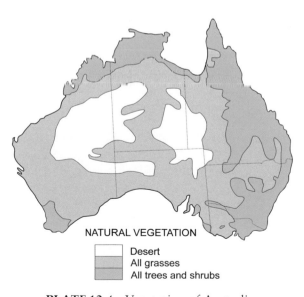

NATURAL VEGETATION

Desert
All grasses
All trees and shrubs

PLATE 12.4. Vegetation of Australia.

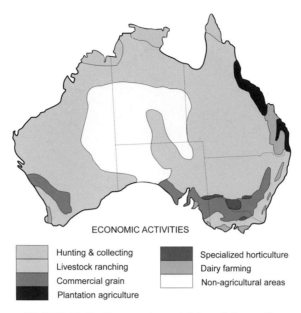

ECONOMIC ACTIVITIES

Hunting & collecting	Specialized horticulture
Livestock ranching	Dairy farming
Commercial grain	Non-agricultural areas
Plantation agriculture	

PLATE 12.5. Economic activities of Australia.

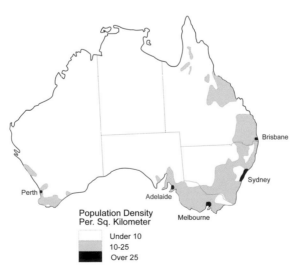

Brisbane

Sydney

Perth

Adelaide

Melbourne

Population Density
Per. Sq. Kilometer

	Under 10
	10-25
	Over 25

PLATE 12.6. Population of Australia.

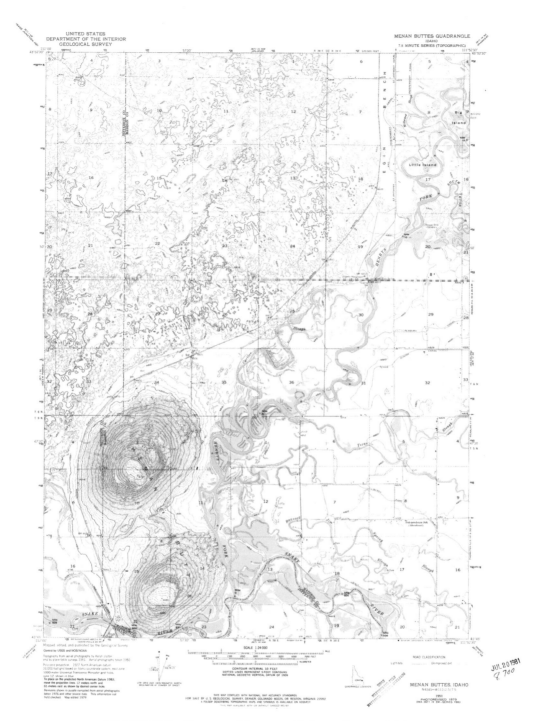

PLATE 12.7. Topographic map of Menan Buttes, Idaho.

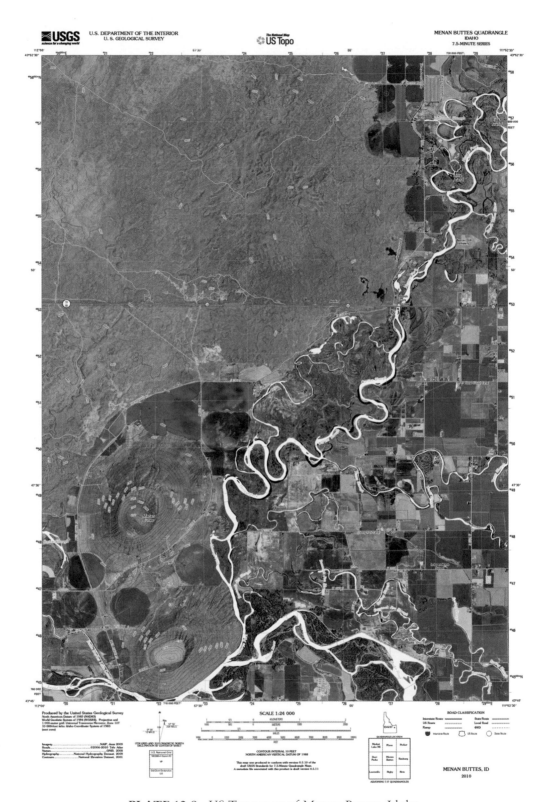

PLATE 12.8. US Topo map of Menan Buttes, Idaho.

PART III

Putting It All Together

CHAPTER 12

Map Interpretation

> . . . in a general sense, maps are information givers,
> and as such they have endlessly varied capabilities.
> No other contrivance of man tells so much about
> so wide a realm in so small a space.
>
> —*Mapping* (p. 74)

Armin K. Lobeck, writing in *Things Maps Don't Tell Us* almost 60 years ago, said, "Maps tell us a great deal. But there are some things which maps don't tell us, interesting things too. By this I mean that there are some facts hidden on the map for us to read if we know how to do so" (1956, p. ix). In previous chapters we have concentrated on map reading and map analysis; in the words of G. H. Drury (1960), this is akin to learning a language, but map interpretation is learning to speak the language. Map interpretation, rather than focusing on determining distances, elevations, locations, and quantities, brings all of these basic tasks together to paint a picture of the nature of an area, its physical and/or human features.

In map interpretation, there are four guidelines to keep in mind:

1. Recognize spatial patterns.
2. Describe patterns.
3. Correlate different patterns.
4. Be aware of cartographic limitations.

We must recognize that in map interpretation we aren't describing lines and colors on a piece of paper or computer screen, but rather we are learning about and describing a real place *from* those lines and colors, not what the map looks like. It isn't enough to say that the contour lines are close together; what does the spacing of the contour lines tell you about the nature of the land, its steepness or flatness?

What do we mean by patterns? First, *pattern* is the arrangement and placement of features. One must be careful of "windmill counting." Finding 20 windmills on a map tells us nothing; because of generalization, there might actually be far more than 20 windmills, but more importantly, so what? What does the number of windmills tell you? What is their pattern; that is, where are they located? Are they in passes? On hilltops? This is more important than the numbers; numbers change frequently, whereas patterns tend to remain static. Pattern recognition is essential in map interpretation.

Second, patterns are not just for cultural features such as roads or houses; natural features also have patterns. We have mentioned stream patterns elsewhere, but, upon closer examination, even apparently random items have patterns. Vegetation is an example. Joshua trees, a desert plant, are found only at elevations between 2,000 and 6,000 feet in the Mojave Desert of southwestern California, Nevada, Utah, and Arizona; in fact, their existence defines that desert. This elevation range is a pattern. Most plants have such locational ranges.

Pattern recognition and correlation go hand in hand. In our Joshua tree example, the vegetation pattern is tied to an elevation or terrain pattern. If we see that population in an area is found along rivers, we are correlating a cultural and a physical pattern. We must be aware that correlation is primarily noting a relationship between two or more patterns, not stating a cause and effect. Correlation is not always causation. There may be factors in play with a correlation that are not obvious on the map, and so additional information may be needed. For example, a thematic map showing U.S. agriculture shows that the patterns of hog production and corn growing are nearly identical (Figures 12.1 and 12.2). The reader will recognize that clearly there is a relationship between the two, but what is that relationship? Is it that hogs are fed corn? Or perhaps that hog fertilizer is used in cornfields? Either assumption will result in the patterns. At this point, the reader needs additional information from other maps or texts.

Finally, we must remember that maps have limitations. To recap the limitations described in Chapter 1, maps are limited by being smaller than reality, largely symbolic, generalized, and sometimes biased. Maps are usually designed for a specific purpose, and no single map can show "everything." In our hog and corn examples, the maps do not show climate, rainfall, natural vegetation, soil quality, terrain, and myriad other factors that might influence the patterns. We also recognize that some maps are deliberately biased or misleading, as in the case of propaganda. In addition, we must remember that we bring our own biases and stereotypes to map interpretation. We may "know" that something is true and may accordingly project that belief onto the map. For example, a map of China's agriculture may show an area as growing grains, and the reader assumes from various stereotypes and beliefs that the grain must be rice, whereas actually much of China raises wheat.

The majority of early map interpretation books and articles focused on topographic and geologic maps and on identifying geologic features—subjects that can be seen on the ground and confirmed in the field. But we have looked at many different

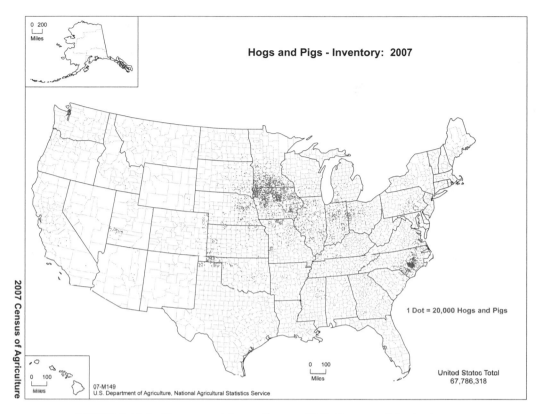

FIGURE 12.1. Dot map of hogs. *Source.* U.S. Census of Agriculture.

types of maps in the preceding chapters that cannot be "field checked." Thematic maps from databases are a case in point.

In the following sections, we will use maps to analyze a variety of subjects; we will progress from a single map showing one subject to multiple maps to examine the nature of an area. To do these interpretations, we will "dissect" the maps to look at the various elements and patterns, and we will develop a structure for interpretation. We will also use *only* the maps for our interpretations to see just how much information is available from maps alone. In a real-world scenario, we would look at other sources as well.

GEOGRAPHY OF CORSICA

Plate 12.1 is a simple map of the island of Corsica and its surroundings; it is similar to the general regional maps found in an atlas. If we examine the map closely, we can find the following patterns: location, both relative and absolute; size and shape of the island; rivers, cities, transportation network. We can also look for relationships between the patterns, that is, correlate the patterns.

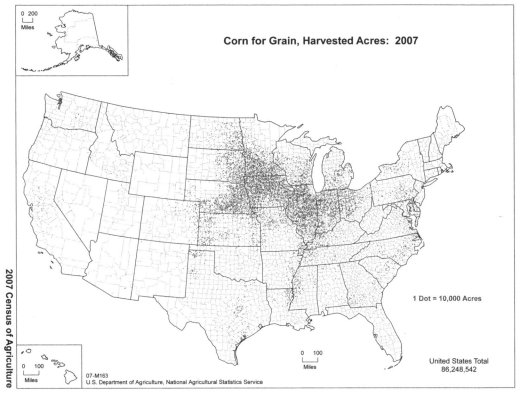

FIGURE 12.2. Dot map of corn. *Source.* U.S. Census of Agriculture.

Size, Shape, Location

The index to an atlas might show Corsica at 42°N and 9°E. However, by looking at the map, we can see that these coordinates are at the center of the island; that doesn't tell us much. It would be better to estimate the island's longitudinal and latitudinal extent from the parallels and meridians. This not only locates the island but also gives an idea of its size. Thus we can say that its latitudinal extent is from 41°N to about 43°N and from about 8°30'E to about 9°30'E. Another useful type of location is *relative location*; that is, where is Corsica with respect to other places? We can see that it is in the Ligurian Sea about 60 miles at its closest point from Italy and over 100 miles from France. It is only 10 miles from Sardinia, another island to its south across the Strait of Bonifacio. We also note that it is a French island, so the distance from France has some significance. The island is roughly oval, with its longest extent roughly 120 miles north–south, and its width at the widest point about 60 miles. We can see a long peninsula extending to the north.

Coastline

The outline of the map shows that the west coast of Corsica is more irregular than the east coast and that the east coast has at least three bays. This could indicate less drainage on the east. We do know that both the east and west sides of the island have at least two rivers.

Terrain

Because this map has no contour lines or other indications of terrain, we cannot describe the terrain in any detail. However, we do see that there is drainage to both the east and west, which would indicate a central highland, and we know that elevations reach over 7,000 feet with Mount Cinto in the north at 8,878 feet and Mount Incudine in the south at 7,001 feet.

Population

The map shows nine cities and the symbols give an idea of the populations, but there are many cautions here. Both Bastia and Ajaccio are shown with a symbol that represents towns with populations between 10,000 and 50,000. This is a considerable range, and from the map alone, we have no way of determining the exact population. Thus, simply saying that the populations of these towns are over 10,000 is the best we can do. We have noted that nine towns are shown; this doesn't mean that there are only nine towns on the island. The cartographer, in the process of generalization, preserves patterns as much as possible. Since probably not all towns can be shown, the cartographer will try to show the general pattern. For Corsica, we can see that the towns are largely coastal, although Corte is located in the interior on a river. From this map we don't know if there are other towns, but we can note that the towns are fairly evenly distributed on all coasts and that the majority have populations under 10,000.

Transportation

Our map only provides some basic information about roads and nothing about railroads. Although Corsica is an island, there is no information on what kinds of ports might be at the various coastal cities. Therefore, the only transportation information we can gain from this map is roads. We can see that a road encircles the coast and joins the coastal towns. We also see that two roads go through the mountains, one joining Calvi on the west coast to the east coast and another trending southward and then west to reach Ajaccio.

We must be aware here of the limitations of our map. Other maps at this same scale might show other features, such as railroads and airports; larger-scale maps would show more detail. But we have been able to gain a great deal of information from this small map, and this could be used as a starting point for a more detailed interpretation that would include larger maps, thematic maps, and written information.

GEOGRAPHY OF AUSTRALIA

Here we will look at a series of maps of Australia showing precipitation, climate, vegetation, economy, and population and integrate that information into a description of Australia. These maps are the kind we can find in the thematic sections of an atlas. The procedure of interpretation is much like that for Corsica except that many more maps are available. These maps can be considered layers of information, and we can analyze them and correlate patterns to get a more complete "picture" of Australia. We will look at each map separately and then combine them to describe the geography of Australia.

Rainfall of Australia

Plate 12.2 is a thematic map showing the average annual rainfall of Australia by means of isohyets—isarthmic lines that join points having the same amount of rainfall. The map shows us only information about average rainfall; it cannot tell us about temperature or elevation.

To interpret the information, we first notice that the isohyet interval is uneven. The lines are 10, 20, 40, and 80 inches; thus, we must insert (either mentally or actually) the "missing" isohyets to make a meaningful interpretation. Sketch in the map isohyets representing 30, 50, 60, 70 to get the complete picture. We could also draw a north–south and an east–west profile to visualize the rainfall pattern and amounts.

The first thing we notice is that the rainfall pattern is asymmetrical; that is, there is more rainfall in the east than in the west and more in the north than in the south. We can also see that the rainfall is highest on the coasts, especially the eastern and northern coasts, and that the interior of the continent receives less than 10 inches of rain annually. When we look at the spacing of the isohyets, we see not only that the rainfall is highest on the coasts, but also that it increases rapidly as one nears the coasts. We also note that the highest rainfall is in the north.

Climate

The map in Plate 12.3 shows six climate regions; other maps might show more or fewer based on the classification system used. Here we see that Australia has a dry interior, with the center being classed as arid. On the northern coast, the climates are tropical with a winter dry season. The southern coasts are humid and temperate, with the south and southeast having a cool summer. The southwest is also humid and temperate but receives rainfall throughout the year. The east coast has a dry season in the winter, and the western coasts are dry.

Vegetation

The map in Plate 12.4 is a very simple map of vegetation that shows only three categories: desert; grasses; trees and shrubs. Again, more categories might be found on other maps. The map shows definite colors separated by black lines. This must not be interpreted as sharp boundaries between vegetation types; such boundaries do not usually occur in nature, and along the "boundary" there is usually intermingling of vegetation. These lines do not have values as the isohyets on the precipitation map do. Also, we must remember that these are *natural* vegetation types and that plants may have been introduced into the areas so that there may be variations. Our map is limited in that it does not show transitions or all kinds of vegetation. What we can learn from this map is that the interior is desert surrounded by areas of grassland. The coasts on the north, southwest, east, and southeast have trees and shrubs. The coasts on the south and west are grassland.

Economic Activities

Plate 12.5 is the most complex of our maps, showing seven categories; again, other maps might show more or fewer. Probably the first thing we notice is that the interior is a nonagricultural area that is surrounded by areas of livestock ranching. We can see areas of commercial grain farming along the southeastern and southwestern coasts and also somewhat inland of those areas. We see that the central east coast is an area of dairy farming and plantation agriculture and that in the southern interior there is some specialized horticulture. The farthest north areas are nonagricultural hunting and collecting areas.

Population

Plate 12.6 shows us that Australia is sparsely populated, with under 10 people per square kilometer, and the majority of the population is located on the southwestern, eastern, and southeastern coasts. Populations of 10–25 people per square kilometer can be found a considerable distance inland in the east and southeast. The most heavily populated areas with populations over 25 people per square kilometer are those around the metropolitan areas.

Thus, when we correlate these patterns, we can see that Australia has an interior desert, but conditions are wetter as one moves to the periphery of the continent. There are essentially no economic activities in the desert area, and population is extremely low. As one moves coastward, there are large areas of grassland that are associated with livestock ranching and some grain production. The population remains low in those areas. The southeastern, southwestern, and eastern coasts are temperate and covered with trees and shrubs. The greatest population density and the majority of the economic activities are found in those areas. The far northern coasts are tropical; have the highest rainfall, although it falls in the summer months; and have a low population density. As important as what the maps tell us for this interpretation is what the maps *don't* tell us. We have already observed that the maps are very generalized, with the vegetation map only showing three categories. Our maps give us no information about terrain, latitude, and longitude. These factors greatly influence climate and, in turn, vegetation and economy. Thus, while we have a general picture of the continent, we have many unanswered questions.

GEOGRAPHY OF MENAN BUTTES, IDAHO

We will interpret the nature of the Menan Buttes area using large-scale topographic maps, both conventional and the newer US Topo maps (Plates 12.7 and 12.8). Because these maps are too large to print here at full size, the reader must order copies or view them online at *www.usgs.gov/pubprod*. However, even with the much reduced versions printed here, we can see some of the patterns. The interpretation that follows is from the full-size maps. As we noted earlier in Chapter 9, thousands of bits of information may be found on any topographic map. Table 8.1 provided some guidelines.

Introductory Information

This is found on the collar or margins of the map. This information locates the study area and also indicates the accuracy and currency of the map. Menan Buttes is located in eastern Idaho; the coordinates given for the corner of the map closest to the equator and prime meridian (SE corner) is 43°45'N and 111°52'30". Since this is a 7.5' series, the map extends 7.5'N and 7.5'W of that location. The scale is 1:24,000, the standard scale for the 7.5' series. If you are making measurements from the reduced image, remember that only the graphic scale is accurate when the map is reduced or enlarged. The contour interval is 10 feet, which allows us to estimate or *interpolate* elevations to half that interval. The area covered is 8.75 miles by 6.5 miles, or 56.875 square miles. This map was originally produced in 1951, and it was photo revised in 1973. These dates are important because it means that information shown on the map is 40 years old. While physical features may not have changed much, cultural or human features are probably considerably different.

Our interpretation will not focus on just physical or cultural features, but rather it will be a geographic analysis that includes both physical and cultural aspects. A useful way of approaching and describing an area is by *regionalization*. *Regions* are areas of relative homogeneity. We can have geologic, drainage, or cultural regions, for example, but here we will define geographic regions that have homogeneity in both physical and cultural features. Menan Buttes is an especially striking area in that four areas stand out immediately: the northwest corner that has irregular terrain; the river valleys; the Menan Buttes for which the quadrangle is named; and the relatively flat area in the east and southeast.

Terrain and Drainage

The irregular area has local relief (the difference between highest and lowest points) of only about 100 feet between 4,850 and 4,950 feet. It covers about 15 square miles on the map, but we must note that it apparently extends beyond the borders of the map to the north and west and without access to the adjoining quads, we don't know how large the area is in total. There are no streams in this area. The area has the appearance of a lava flow. The two buttes are roughly conical with central craters, as we can see from the depression contours. The northern butte rises from an elevation of 4,850 feet to 5,500 feet, and the southern butte rises to 5,300 feet. The northern crater is about 300 feet deep and the southern about 200 feet. The drainage is a combination of radial and centripetal. Although there are no permanent streams on the buttes, the shape of the contours indicates drainage into the craters from the rims and outward along the flanks. Both features are steeper on the southwest than the northeast. These features then are not actually "buttes" but volcanic features. We can't determine precisely what type of material from these maps. The river valleys follow the meandering Snake River and Henry's Fork of the Snake. The rivers are meandering with old meander scars (seen from the contour lines) and islands. Henry's Fork flows southward and joins the Snake that flows westward. We determine the direction of flow from the angles of intersection of the rivers and the shape of the islands; the narrowest angle points downstream. The Snake River has levees on both sides, probably for flood control. The eastern region is essentially flat, with some tributaries of the major rivers flowing through it.

Cultures

Both roads and railroads are seen on this quadrangle, and their patterns vary by region. The single railroad through the quadrangle follows the terrain, skirting the irregular area to the east and bending around the buttes. The irregular region had only a few unimproved dirt roads in 1951 that tended to follow the contour of the land, but a new numbered road, state route 33, was built before 1979. It goes due west from Rexford (off the map in the adjoining eastern quadrangle) to join Interstate 15, a total of 12 miles to the west of the edge of the quad. There are no roads on the buttes and few in the river valley. Roads in the flatlands roughly follow a rectangular pattern. Since this is a PLSS area, this pattern would be expected. The area covers parts of four townships, with T6N and T5N meeting in the southern third of the area. Their contact is a correction line, and we can see that the county line, the section lines, and many of the roads have a distinctive "jog" along the correction line.

There are no towns on the quadrangle, but the map indicates that Rexburg is on the quadrangle to the east. Neither the irregular region nor the butte and river regions have houses of any sort. The flatlands have scattered houses along the roads. Because USGS topographic maps do not show crops except for orchards and vineyards (as many European topographic maps do), we cannot determine what, if anything, is grown in the area from just the map. Because of the number of houses and their spacing, we can guess that they are farms.

When we look at the US Topo map, we can add to this information. This map has the same scale and coverage as the older topographic map, but it was produced in 2010 and is therefore more up to date, especially for cultural features. Because it is on an orthophoto base, we can find more information about agriculture and population.

US Topo has several advantages for our purposes. These maps are easier to update than the older topographic maps, so new information can be added more quickly. The orthophoto is in color, so we can see areas of vegetation and crops. Because it is made of digital layers, if we are viewing on a computer monitor, we can focus on one pattern at a time and view only that pattern. We can zoom in and out of the map to increase the amount of detail we can see. A disadvantage at this time is that such features as mines and oil wells are not labeled. Future maps will have this information.

Looking at the US Topo map, we can see that the basic patterns are the same, but we have a better idea of the vegetation. The irregular area is covered in grasses and some small shrubs; the buttes also have this vegetation. The river areas are much greener, and there are trees. The eastern areas are covered with crops; little or no natural vegetation remains.

We can see that the number of roads has increased, although the basic patterns remain. The railroad no longer exists, but a road follows its former path, and traces of its former route can be seen. A road now leads to the top of the southern butte. There are still no towns or other settlements within the boundaries of the quadrangle, although there are some clusters of farms and houses that might constitute villages. On this map we can see the size of the farmsteads and the types of buildings. These are all concentrated in the eastern region.

The irregular region has several cleared areas, with dirt roads leading to them. These could be extractive industries such as mining or drilling, or even grazing areas, but without further information we don't know. We can't determine this from the maps. The eastern region is clearly agricultural. We can see rectangular fields with row crops and large circular green areas. The circular areas, which are quite distinctive when seen from the air, are fields irrigated by center pivot irrigation. These are giant sprinkler systems that rotate around the center pivot and irrigate crops in dry areas. The crops are planted in a circular pattern. The size of fields irrigated by these systems varies, but the largest fields have a radius of about 1,600 feet. The fields near the Buttes are about this size, so these are large operations. The flatlands are marked largely by row crops in linear patterns. Because there are a number of canals in the area, we can speculate that these crops are also irrigated.

Thus, we can describe the Menan Buttes quadrangle as having four distinct regions: the irregular northwest, the flat east, the Buttes, and the river valleys. The total area, 56.875 square miles, is divided according to the U.S. Public Land Survey System. It includes two counties, Jefferson on the west and Madison on the east. No towns or cities are located within the quadrangle, but Rexford is 6 miles to the east of the quad. There is a state highway that goes straight in an east–west direction across the middle of the area; it leads to Interstate 15, which is 12 miles from the western edge of the quadrangle. The northwest can be characterized as an area of irregular terrain and probably a lava flow, with grasses and small shrubs. It is an area with few roads, and the economic activities are probably extractive industries and perhaps grazing. The Buttes area is made up of two asymmetrical volcanic cones on a NE–SW axis with depressions at their tops and steeper slopes on the southwest. Agriculture in the flat areas surrounding the cones to the west of Henry's Fork is supported by center pivot irrigation. The river region is the floodplain of the Henry's Fork and Snake rivers and is covered with woodland. Levees follow parts of the Snake River. The eastern flatlands are marked by rectangular roads and a series of canals. This region is agricultural and is likely irrigated.

We have learned a great deal about this area from just the maps, but questions still remain. These questions could be answered by field work, that is, traveling to Menan Buttes and seeing it from the ground and by consulting other maps, such as geologic maps, population maps, and economic maps.

FURTHER READING

Drury, George H. (1960). *Map Interpretation* (2nd ed.). London: Pitman.
Keates, J. S. (1982). *Understanding Maps*. New York: Halsted Press.

Kimmerling, Jon, Aileen Buckley, Phillip Muehrcke, and Juliana Muehrcke. (2010). *Map Use: Reading, Analysis and Interpretation* (7th ed.). Redlands, CA: ESRI Press.

Lobeck, Armin K. (1956). *Things Maps Don't Tell Us*. New York: Macmillan.

Pearce, Margaret, and Owen Dwyer. (2010). *Exploring Human Geography with Maps* (2nd ed.). New York: Freeman.

CHAPTER 13

Epilogue: The Future
of Maps and Map Reading

Today with speeded-up transportation, swift
mechanized warfare, messages delivered within
seconds instead of months or years, map changes
are coming in faster than ever before.

—*Down to Earth* (p. 2)

When David Greenhood wrote *Down to Earth, Mapping for Everybody* in 1944, he envisioned looking at earth from the moon, but the maps he described were conventional maps drawn with pen and ink and printed on paper. The tools used were much the same as those used for centuries. The list of inventions related to maps and map use in the 70-year period since 1944 is jaw-dropping: space flight and satellites, GPS, GIS, digital photography, the Internet, personal computers, drones, DVD players and recorders, photocopying machines, high-quality printers, three-dimensional printers, tablet computers, digital readouts, and smart phones. Perhaps most readers have no memory of using a film camera and waiting for pictures to be developed before showing them to friends rather than looking instantly at our phone or camera screen. To older readers these developments were once the stuff of science fiction and comic books.

In this book we have looked at some of the new map types and maps created with the new technology; many technologies such as GPS are now commonplace, and we wonder how we found our way without following the directions on our smart phone. With touch-screen computers, tablets, and phones, we can already change the scale of a map on screen by "pinching" or "stretching" the map, and we can pan across the landscape to view our route as an image or map or combination. We can speak our destination to our phone GPS, and we will be given spoken turn-by-turn directions as well as the route shown on screen. We have seen that we can touch the screen for some maps, and information about the location will be displayed or spoken. Animated maps can take us on a trip through time or space. What is next on the horizon of maps and mapping? (See Figure 13.1.)

It is always risky to attempt to describe the future. Predictions in the 1960s had Americans driving small flying automobiles. Instead, at the beginning of the 21st century, one of the most popular car types was the large sport utility vehicle, the road-bound SUV, no flying cars. Despite the risks of prediction, beginning in the 1980s a spate of articles and book chapters appeared on the future of maps in the new coming millennium. Now that the millennium is here, articles, websites, and blogs are predicting new directions for maps and mapping. Here are some of the predictions and concerns.

One of the first questions asked when speaking of maps of the future is, "Aren't maps obsolete?" Given that Google Maps and other mapping websites are among the most frequently downloaded sites, the quick answer is no. People still need assistance with navigating streets and roads. The next question then is, "Aren't paper maps already obsolete?" This is a more difficult question. Paper road maps are no longer displayed prominently in gasoline stations, state mapping agencies are printing fewer maps, and even the AAA has closed some of its map production facilities. We noted the almost complete demise of street atlases in Chapter 10. However, many users still print out the map and directions from Google to have a map in hand. A paper map is invaluable on a hike in a deep canyon where no satellite signals are available, and navigation maps on GPS and online maps do not show the "big picture," so paper maps are still of use. Whether they will remain useful is a matter of some debate.

In Chapter 3 we talked about revolutions in cartography and observed that they are precipitated when three factors intersect: changes and increases in data, changes

FIGURE 13.1. Word cloud of this chapter.

in technology, and changes in the way we think about maps. We noted that such a revolution had its beginnings in the 1960s, and it continues today at an accelerated pace. At the end of the 20th century and the beginning of the 21st, three factors have been driving the current revolution: (1) an astronomical increase in the amount of data available, (2) rapidly expanding technology (computers, satellites, Internet), and (3) changes in how we think about maps and mapping—a map is no longer confined to being "a graphic representation of the earth's surface drawn to scale upon a plane." While maps of the moon and the heavens have been made for centuries, we now can produce geologic maps of the moon and terrain maps of Mars. Indeed, we are mapping the universe. New representations are based on imagery from satellites and the orbiting Hubble telescope. The *Voyager* mission traveling beyond the solar system will send data from far out in space.

Maps are no longer static representations, but are increasingly dynamic and interactive. The method of map delivery is changing rapidly even for "conventional" static maps. Just as e-readers and e-books have become more common, so have maps that can be read on tablets and e-readers. The technology has especially changed how we think about and define maps. It has blurred the line between mapmaker and map user with open-source mapping software, such as OpenStreetMap, and with data readily available online. Mapmaking is no longer the exclusive provenance of professional cartographers. More and more, map users are mapmakers.

In 1997, D. R. Fraser Taylor introduced the term *cybercartography* to describe the changes in cartography brought about by new technology. The definition of this new term was "the organization, presentation, analysis and communication of spatially referenced information on a wide variety of topics of interest and use to society in an interactive, dynamic, multimedia, multi-sensory format with the use of multimedia and multimodal interfaces" (Taylor, 2003, p. 405). In simpler terms, cybercartography involved using computers, and especially the Internet, to create multimedia and interactive maps. Most new map types fit this definition.

Static maps and atlases can, of course, be viewed on e-readers, and already a large number of atlases are available, many aimed at children. But with the popularity of tablets with color and touch screens, more interactive atlases are possible, not as imitations of maps in paper atlases, but maps that can be queried and combined. Increasingly, cartographers are creating *webmaps*. The U.S. Geological Survey allows users to create simple maps from national data with the National Atlas online. For US Topo, we have seen how the user can turn layers on and off to view only roads or only contours. While these products are not yet truly interactive, they are first steps.

One popular modern map type is the so-called three-dimensional (3-D) map, but although the maps show elevation as well as position, these representations are not truly three dimensional; they are perspective or bird's-eye views, a technique that was popular in the 19th century. Aerial photographs also provided such views. Newer versions are animated visualizations that show terrain features and allow the viewer to "fly through" the landscape; Google Earth spins the world and gives an impression of altitude. At some public venues such as museums, large screens are installed that allow viewers to "fly" to Paris, the Grand Canyon, and other selected sites, interactively through Google Earth. The viewer uses a trackball-type device to pan and soar and even land on the runway at Orly Airport. But again these aren't truly 3-D views

such as a stereoscope or a 3-D film would provide, but they do give the sense of elevation and flying. However, there are many who are working on immersive 3-D maps
projected on a screen, or even as a hologram, that allows the user to "walk through"
the landscape, perhaps to plan a hike, and practice it before setting boot to trail.
Developments in 3-D printing can lead to models of the landscape.

The majority of the maps discussed thus far focus on wayfinding and representations of the land, but increasingly, thematic maps are becoming the subject of
new techniques. Here is where the greatest challenges and changes in maps can be
found. Data drives thematic mapping, and the amount of data generated in the last
decade has been huge. Many of these data are readily available through the Internet.
Whereas once a mapmaker had to go to a library to access census data, they are now
readily available at the Census Bureau website, literally at the cartographer's fingertips. While many of the new maps are made from information from the censuses
and satellites, users are creating and disseminating data through GPS-enabled smart
phones and digital cameras and open-source data sites. In a simple form, it is possible to geotag photos from a GPS-enabled camera and attach the images to a map
on the web to illustrate the road trip the maker took on a summer vacation. In more
complex situations, web users can create a kind of "flash mob" and generate data
that are mapped. As noted above, the distinction between mapmaker and map user
has been blurred, with the user having access to data and the ability to generate maps
through use of Google maps, Google Earth, and the like. This kind of "citizen" mapping has generated a new term, *neocartography*, which refers to mapping on the web
commonly by nonprofessional cartographers using open-source software and data.
The maps may be for the maker's own use or may be disseminated through the web
to millions of viewers. The term *prosumer* has been used for these makers/users. The
International Cartographic Association (ICA) proposed a Commission on Neocartography in 2010, which is now functioning. Neocartography operates through what
has been called Web 2.0, which describes websites that go beyond static representations and utilize animation and sound.

There is no accepted formal definition of neocartography, although according to
the ICA's neocartography section of its website:

> The term neocartographers is being used to describe map makers who may
> not have come from traditional mapping backgrounds, and are frequently using
> open data and open source mapping tools. . . . The availability of data and tools
> allows neocartographers to make their own maps, show what they want, and
> often be the intended audience as well—that is to say they may make the maps
> for themselves, just because they can. (*http://neocartography.icaci.org/mission-
> and-aims*)

The emphasis of neocartography is on the "democratization" of cartography; of
maps being made by nonprofessionals who aren't members of what is seen as the elite
priesthood of cartography. Of course, maps have always been made by amateurs.
Anyone with a pen or pencil can sketch a map on a napkin, or with simple equipment,
such as a ruler or protractor, make a publishable map. Probably the major difference
now is the amount of data available and the ability to disseminate the map through
websites and blogs to millions of people.

LIMITATIONS, CONCERNS, CAVEATS

Since this book focuses on map reading and interpretation, not mapmaking, we can ask, what are the impacts of this cartographic revolution on map use? How will interpretation change? Will it? The new map types, like the old, have limitations. What will it mean to have users creating maps? Here we will examine these limitations, concerns, and caveats.

All new technology goes through the "gee whiz period" of "look what I can do." Later, the value and usefulness of the new products are analyzed. For some map types such as Google Maps and other on-screen mapping sites, their usefulness is widely accepted, and certainly the value of maps on GPS is not disputed. But the usefulness of many other non-navigational or location maps has yet to be evaluated.

All maps have limitations, and webmaps using multimedia are no exception. A choropleth map or a dot map has the same properties when viewed on a screen as when viewed on paper. The guidelines for reading such maps remain the same. We noted in Chapter 7, with regard to dynamic and interactive maps, that maps are only as good as their data and their makers, and here lies the concern for many of the new map types.

One of the major uses of maps is decision making: Where should the new road go? Is this new housing development on an earthquake fault? Where should money be spent for a new school? Is this a proper place for a chemical plant? What route should I take? Decisions on these questions assume that the maps are accurate and not biased.

We have already seen the problems that arise when information on a GPS is incorrect. This is the case for webmaps as well. A recent example involved a Scottish island that had been omitted from Google Maps. Other examples involve disputed boundaries and names. This has been a problem with static paper maps and remains a problem for webmaps. Wars have been fought over a misplaced boundary line.

Data sources may be biased; just because several thousand people are providing data doesn't mean those data are accurate or unbiased. In fact, there is an inherent bias because the contributors come from a specific group, because of either interest or location and because they have access to the web and know how to use it. Web use may seem universal, but it is not yet so.

Neocartographers are not always aware of or concerned with the conventions of mapping. This is often praised as an advantage, but it can also mean that misleading maps are being produced through inappropriate projection and symbol choices. This is not to say that there is only one "correct" choice for any given map, but as we saw in Chapter 5, a distribution map that is created on a nonequal-area projection will give an erroneous idea of the density of the distribution. Books on making maps for the web may contain a brief discussion of projections, but the discussion rarely goes beyond cylinders, cones, and planes. The cartographer may flout the "rules" of cartography for dramatic effect, but may also simply not be aware of conventions that have been used for centuries and that map readers assume to be true. When we are viewing maps on the web, we must remember that maps, like any other information on the Internet, must be carefully evaluated; just because a map is on the web and is seen by millions of people doesn't guarantee accuracy or truth. Therefore the reader must be aware and alert.

The noted geographer and historian of cartography Norman Thrower (2008, p. 1) wrote, "A knowledge of maps and their contents is not automatic; it must be learned, and it is important for educated people to know about maps even though they may not be called upon to make them." It is essential, in our rapidly globalizing world, to heed Thrower's words. The world of maps is expanding rapidly, and with that expansion comes a greater need—and responsibility—to know about maps, why they were made, what they can and cannot show, and what their limitations are. For the future, as for the present, the key to that world is to be an informed, intelligent, and cautious map user.

FURTHER READING

Abrams, Janet, and Peter Hall (Eds.). (2006). *Elsewhere: Mapping New Cartographies of Networks and Territories.* Minneapolis: University of Minnesota Design Institute.

Crampton, Jeremy W. (2010). *Mapping: A Critical Introduction to Cartography and GIS.* Chichester, UK: Wiley-Blackwell.

Frangeš, Stanislav, Nedjeljko Frančula, and Miljenko Lapaine. (2002). "The Future of Cartography." *Kartografia i Geoinformacije, 1*(1), 6–21.

Gibson, Rich, and Schuuler Erie. (2006). *Google Maps Hacks.* Sebastapol, CA: O'Reilly Media.

Jacobs, Ryan. (2013, July 26). "A Lost Scottish Island, George Orwell, and the Future of Maps." *The Atlantic.* Retrieved September 23, 2013, from *www.theatlantic.com/international/archive/2013/07.*

Kraak, Menno-Jan. (2011). "Is There a Need for Neo-Cartography?" *Cartography and Geographic Information Science, 38*(2), 73–78.

Kraak, Menno-Jan, and Allan Brown. (2001). *Web Cartography.* New York: Taylor & Francis.

Mitchell, Tyler. (2005). *Web Mapping Illustrated.* Sebastapol, CA: O'Reilly Media.

Monmonier, Mark S. (1985). *Technological Transition in Cartography.* Madison: University of Wisconsin Press.

Rhind, D. W., and D. R. F. Taylor (Eds.). (1989). *Cartography Past, Present and Future.* London: Elsevier.

Taylor, D. R. Fraser. (2005). *Cybercartography: Theory and Practice.* Amsterdam: Elsevier.

Taylor, D. R. Fraser, and Stephanie Pyne. (2010). "The History and Development of the Theory and Practice of Cybercartography." *International Journal of Digital Earth, 3*(1), 2–15.

Tucker, Patrick. (2013). "Mapping the Future with Big Data." *The Futurist, 46*(4). Retrieved from *www.wfs.org/futurist/2013-issues-futurist/july-august-2013-vol-47-no-4.*

RESOURCES

International Cartographic Association, Commission on Neocartography
http://neocartography.icaci.org

U.S. National Atlas
http://nationalatlas.gov

Appendices

APPENDIX A

Glossary

Active system. A remote sensing system that emits its own electromagnetic radiation, such as radar, or that measures electromagnetic radiation reflected from the surface of an object but not emitted by the object.

Aeronautical chart. A map used by pilots for navigation.

Aliquot part. The standard subdivisions of a PLSS section, such as a half-section, quarter-section, or quarter-quarter sections.

Altitude tint. *See* Layer tint.

Animated maps. Maps that create the illusion of change, either temporal or spatial, by rapidly displaying a series of single frames (Peterson, 1995).

Antarctic Circle. The parallel located at 66½° south of the equator or 23½° north of the south pole. The sun's rays are tangent at the solstices.

Arctic Circle. The parallel located at 66½° north of the equator or 23½° south of the north pole.

Arpent. A unit of land measurement in the French Long Lots survey system, found in Louisiana and other French-settled areas.

Atlas. A collection of maps, bound or boxed, that conform to a uniform format.

Azimuth (magnetic). A spherical angle formed by a true north line (a meridian) and a line that passes through the observer and the object observed measured clockwise from north. The spherical angle between any great circle and a meridian.

Azimuthal projection. A class of projections that shows azimuths correctly from the center point; great circles through the center point are straight lines. Also called a zenithal projection.

Bar scale. *See* Graphic scale.

Base line. The principal east–west line of rectangular survey (PLSS) of the United States. It meets its corresponding principal meridian and the origin or initial point.

Bearing. Measurement of direction between two points. The compass is divided into four quadrants.

Cadastral maps. Maps that show property boundaries.

Cardinal direction. The compass directions of north, south, east, and west.

Cartogram. An abstract and simplified map for which the base is not true to geographic scale. In the most common forms, the areas of features are drawn according to a value, such as population, or a time scale is used instead of a distance scale. *See also* Linear cartogram; Value-by-area cartogram.

Cartography. The art, science, and technology of making maps and their study as scientific documents and works of art.

Choropleth map. A type of quantitative map on which statistical or administrative areas are shaded proportionally to the value represented.

Collar. As used by USGS, the collar of a topographic map is the space that surrounds the map proper. Also called the margin.

Compromise projection. Projection that has no special properties, but usually has a good appearance.

Condensed projection. Projection having a portion of the grid removed to permit a larger scale map to be presented on a given page size.

Conformal projection. Projection on which the shapes of very small areas are preserved. Parallels and meridians cross at right angles, and scale is the same in every direction about a point.

Conic projection. Projection that appears to have been projected onto a cone.

Contour interval. The vertical distance between two adjacent contours.

Contour line. A line that joins all points having the same elevation above or below a datum, usually mean sea level. A type of isarithmic line, also called isohypse.

Contour spacing. The horizontal or map distance between two adjacent contour lines.

Correction lines. Standard parallels on the U.S. Public Land Survey System. These lines compensate for the convergence of meridians toward the poles.

Cylindrical projection. Projection that appears to have been created by projecting the earth's grid onto a cylinder.

Data point. A point for which statistical data are available.

Depression contour. A contour line with hatch marks on it used to indicate a closed depression. The hatch marks point downward.

Depth sounding. Depths beneath the ocean, lake, or river.

Developable surface. A simple geometric form that can be cut and flattened without distortion. The three developable surfaces are the cylinder, the cone, and the plane.

Digital elevation model. There is at this time no uniformly agreed upon definition, but basically it is a digital model of terrain of the earth or another planet created from elevation data.

Direction. The location of one point in space with respect to another.

Dot density map. A variation of the simple dot map that places dots randomly within the enumeration area.

Dot distribution map. A dot map on which the dots are placed in their area of occurrence within the enumeration area.

Dot map. A representation of geographic phenomena on which dots represent a specified number of the phenomena being mapped.

Dynamic maps. Maps on the computer that are interactive or animated. They may have sound effects or clickable pictures.

Electromagnetic energy. All energy that moves with the speed of light.

Electromagnetic spectrum. The ordered array of known electromagnetic radiations from cosmic rays at the short-wavelength end of the spectrum to radio waves at the long-wavelength end of the spectrum.

Ellipsoid. A mathematical figure that approximates the shape of the earth in form and size; it is used as a reference surface for geodetic surveys.

Equal-area projection. A map projection that preserves a uniform-area scale; countries, continents, and other areas maintain their correct size with relation to one another. Also called an equivalent projection.

Equator. An imaginary great circle drawn around the surface of the earth midway between the north and south poles. It divides the earth into two equal hemispheres and is designated as 0° latitude.

Equidistant projection. A projection that shows distance correctly along certain lines or from certain points.

Flow line. A linear symbol whose width varies in direct proportion to the quantity represented.

Form lines. Approximate contour lines that show the shape of the land but are "sketched in," not drawn from measurements.

French Long Lots. A cadastral system used by the French, also called the seigneurial system, in which the lots front on a river and are long and narrow.

Frequency. The number of waves that pass a point per unit time.

Generalization. A basic cartographic procedure that reduces the amount of information presented in order to create a clearer communication. Because maps are drawn smaller than reality, they must be generalized.

Geographic information system (GIS). A computer-based system for collecting, managing, analyzing, modeling, and presenting geographic data for a wide range of applications.

Geologic map. A map that shows the geologic features of an area, including such things as rock types and fault lines.

Global positioning system (GPS). A satellite-based navigation system based on 24 satellites placed into orbit by the U.S. Department of Defense. The satellites allow a receiver to calculate latitude, longitude, and elevation as well as speed and direction.

Gradient. The inclination of a linear feature to the horizontal.

Graduated circle. A circular symbol that is drawn so that its area or apparent area is proportional to the amount represented; also called a proportional circle.

Graphic scale. A graduated line marked in ground units that allow distances to be measured from a map. Also called a bar scale.

Graphic variables. The graphic characteristics of symbols. These include shape, size, tone, hue, orientation, and pattern. Also called visual variables. On multimedia maps, these can also include sound, touch, and duration.

Graticule. The system of parallels and meridians on the earth or globe.

Great circle. A circle on the earth's surface formed by a plane that passes through the earth's center and bisects it into equal hemispheres. The shortest distance between two points on the earth's surface is along a great circle arc. Also called an orthodrome.

Grid north. A navigational term that refers to the northward direction along the grid lines of a map projection. It is not the same as true north, the direction of the north pole, or magnetic north, the direction of the magnetic north pole. Some maps include all three norths.

Guide meridian. A part of the U.S. Public Land Survey System. A meridian drawn between the principal meridians.

Hachure. Short, straight lines drawn in the direction of a slope to indicate relief.

Haptic maps. Haptic pertains to touch. Haptic maps use force-feedback devices that allow the user to feel map features.

High oblique. An image that shows the horizon.

Historic map. A map that was made in the past.

Historical map. A modern map that represents events and places of a historic time, such as modern maps of the American Civil War.

Hypsometric tint. Color applied to the area between two selected contours. It is a method of showing relief on maps. Commonly, cool colors are used for low elevations and warm colors for high elevations. *See also* Altitude tint; Layer tint.

Imagery. Representations of part of the earth or other body created with an optical or electronic apparatus, such as a camera.

Index contour. A numbered contour line on a topographic map. These are usually every fifth contour and make the map easier to read.

Inset map. A separate map located within the neat line of a larger map. It may be a portion of the map at an enlarged scale, an area that falls outside the borders but included within for convenience, or a smaller scale map of surrounding areas included for location purposes.

Instrument chart. An aeronautical chart used when flying with instrument flight rules as opposed to visual flight rules.

Interactive map. A map that allows the user to click on parts of the map and interact with various features.

Intermediate contour. Unnumbered contour lines on a topographic map. The difference in elevation between two adjacent contours is the contour interval of the map.

International Date Line. An imaginary line on the earth opposite the Greenwich Prime Meridian. It roughly follows the 180° meridian, and the date changes at this line. A person traveling eastward across the line loses a day; a person going westward gains a day.

Interpolation. Estimating values from an isarithmic map.

Interrupted projection. A projection on which the central meridian is repeated in order to reduce distortion. It permits each area (usually a continent) to be placed in the zone of best representation.

Interstate highway. Properly, the Dwight D. Eisenhower System of Interstate and Defense Highways, which is a network of limited-access highways that are part of the national highway system of the United States.

Isarithm. A line that joins all points having the same value above or below some datum. Also called an isoline or isogram.

Isobath. A line that joins all points the same depth below a datum, usually sea level.

Latitude. Angular distance north or south of the equator; measured from 0° to 90°.

Layer tint. The use of colors between different isarithmic lines. When used with elevation contours, the method is called hypsometric tints or altitude tints.

Legend. An explanation of the symbols and conventions used on a map.

Linear cartogram. A cartogram concerned with distances.

Linear scale. *See* Graphic scale.

Longitude. Angular distance east or west of the prime meridian; measured from 0° to 180°.

Low oblique. An oblique photo that does not show the horizon.

Loxodrome. *See* Rhumb line.

Magnetic north. The direction of the north-seeking end of a magnetic compass needle. It is not the same as true north. The magnetic north pole is not at the same place as the geographic north pole.

Map. Traditionally defined as "a graphic representation of all or a part of the earth or some other body, usually to scale and on a plane," but because some representations that we recognize as maps are not drawn to exact scales or are not graphic, a map can be defined as a spatial representation of information.

Map analysis. Involves calculations from maps, such as determining slope and gradient, computing areas, and drawing profiles.

Map interpretation. The highest level of map use that may involve more than one map to determine the nature of an area or a distribution. It requires recognizing spatial patterns, correlating patterns, and bringing all of the available map information together to describe a place or subject.

Map reading. The most basic aspect of map use: finding locations, recognizing symbols and what they stand for, and basic wayfinding.

Margin. The area of a map sheet surrounding the map information. On US Topo, this area is called the Collar.

Mathematical projection. A projection that can only be constructed from mathematical formulas rather than by geometrically projecting onto a developable surface.

Meridian. A line that connects all points that have the same longitude. Meridians are great circles and converge at the poles.

Metes and bounds. An unsystematic method of surveying land for property ownership. Originally based on natural features, it results in a "crazy-quilt" pattern of property parcels.

Multimedia maps. Maps that use multiple types of images such as illustrations, graphs, and text; with computers, these are expanded to include animation, video, and sound.

Multisensory maps. Maps on a computer that utilize sight, touch, and sound as symbols.

Nautical chart. Also called hydrographic chart. A chart designed for navigation at sea or other waterways.

Nautical mile. A unit of distance equal to 1 minute of arc along a great circle or 1,852 meters or 6,076.1 feet.

Neat line. A line bounding the detail of the map.

Neocartography. A new term that refers to creating maps on the Internet, often interactive and made with open-source data by nonprofessionals.

Nominal scale. (1) The scale used on a map. It is true only along certain lines or from certain points owing to projection distortions. (2) The scale used for an unrectified aerial photograph or image. It is true only for certain areas.

Oblique photo. An aerial photograph taken with the camera's axis at an angle other than perpendicular to the ground.

Orientation. Establishing a relationship in direction with the points of the compass.

Parallel. A line that joins all points having the same latitude. Parallels are true east–west lines that encircle the globe. Only the equator is a great circle.

Passive system. A remote sensing system that records energy reflected from the object being viewed.

Persuasive cartography. A type of cartography whose main object or effect is to change or in some way influence the reader's opinion. Advertising maps and propaganda maps are both examples of persuasive maps.

Photogrammetry. The science of obtaining accurate measurements from photographs; mapping or surveying by photographic methods.

Photorevision. On older USGS topographic maps, revisions were made by use of air photo imagery and added to the map without field checking the information. These additions were usually shown in purple.

Pie chart. A circular symbol divided into sectors to indicate proportions of a total value. Often combined with proportional circles. Also called pie graph, segmented circle, or sectored circles.

Plane of the ecliptic. The plane of the earth's orbit around the sun. The earth's equator is tilted 23½° to the plane of the ecliptic.

Plane projection. A transformation of the earth's grid onto a plane surface.

Planimetrically correct. Features are shown in their correct horizontal positions.

Platform. In remote sensing, the platform is the device that carries the sensor. Kites, airplanes, and satellites are examples of platforms.

PLSS. Public Land Survey System. A systematic, rectangular cadastral survey system employed in the United States.

Portolan chart. A type of early chart used to aid in sea navigation. Typically, rhumb lines radiating from compass roses criss-cross the chart.

Prime meridian. The meridian adopted as the origin (0°) for measurement of longitude. The meridian through Greenwich, England.

Principal meridian. A starting meridian for a portion of the U.S. Public Land Survey System. There are 31 principal meridians in the United States.

Profile. A cross section of part of the earth's surface created by plotting elevations from a contour map along a linear traverse. The horizontal scale represents linear distance along the traverse, and the vertical scale represents elevation. The vertical scale is commonly exaggerated to bring out variations in terrain.

Projection. A systematic arrangement of all or a part of the earth's spherical grid upon a plane.

Propaganda map. A map designed to persuade or influence the reader; the connotation of propaganda is usually negative or untruthful.

Proportional symbol. A point symbol, such as a circle or square, that is drawn so that its area is actually or visually proportional to the amount represented.

Quadrangle. A name for any USGS topographic map.

Qualitative thematic map. A thematic map that shows nominal information, that is, some quality of the data as opposed to numerical information.

Quantitative thematic map. A thematic map that uses numerical data.

Range. A vertical column of townships in the U.S. Public Land Survey System.

Rectified. An image that has been transformed so that the features shown are in their correct planimetric positions.

Region. An area of relative homogeneity or well-defined functional interaction.

Relief shading. A method of creating a three-dimensional effect on maps to simulate an impression of the terrain.

Remote sensing. Detection and/or recording of data about an object without being in physical contact with the object.

Representative fraction. The scale of a map expressed as a fraction or ratio that relates distance on the map to distance on the ground in the same units. Also called natural scale or fractional scale.

Rhumb line. A line that cuts all meridians at a constant angle. Also called a line of constant compass direction, line of constant bearing, or loxodrome.

Rise. When calculating slope or gradient, the rise is the difference in elevation between the starting and ending points of the slope/gradient.

Run. When calculating slope or gradient, the run is the distance between the starting and ending points of the slope/gradient. For calculations, run and rise are expressed in the same units.

Scale. The ratio of the distances on a map, globe, model, or profile to the actual distances they represent.

Seigneurial system. A property system used in French-settled areas of the United States that is typified by long, narrow parcels of land. Also called French Long Lots.

Sensor. An instrument designed to gather information about the environment that is not in contact with the environment. The camera was the earliest sensor used in remote sensing, but radar, sonar, and other tools are also used.

Slope. The inclination of a surface with respect to the horizontal.

Small circle. A circle on the earth's surface whose plane does not pass through the earth's center.

Software. Programs and data files for a computer. Programs provide instructions to the computer, whereas data files provide information.

Sound map. A map that uses sound as a symbol.

Standard parallel. (1) A parallel on a conic projection that has no distortion. It is the line where the assumed developable surface touches the generating globe. (2) In the U.S. Public Land Survey System, one of a series of parallels that are 6 miles north or south of the base line.

State Plane Coordinate System. A series of geographic zones for the United States, each of which is a coordinate system.

Static map. Noninteractive map viewed on a computer.

Supplemental contour. A contour line added between the normal contours of a map in areas where the elevation change is too small to be shown with the stated contour interval.

Tactile map. A map used by the visually impaired on which all symbols are raised and lettering is in Braille.

Temporal map. A map that shows time.

Thematic map. A map that features a single distribution, concept, or relationship and for which the base data serve only as a framework to locate the distribution being mapped.

Time zone map. A map that shows the time zones for the world or a country.

Topographic map. A large-scale, general-purpose map that shows elevation, usually in the form of contour lines, as well as cultural features such as transportation and economic characteristics.

Topography. The characteristics of the earth's surface, including physical features, such as relief, and cultural features, such as settlement and economy.

Township, civil. A minor civil division smaller than a county; a form of local government. Widely used in New England. The term is not synonymous with survey township, although in some areas the two may coincide.

Township, survey. An approximately 6-mile-square parcel of land that contains 36 sections. Also used to refer to a horizontal row of townships.

Tropic of Cancer. The parallel located at 23½° north of the equator. It marks the farthest north that the sun's rays strike the earth perpendicularly.

Tropic of Capricorn. The parallel located at 23½° south of the equator. It marks the farthest south that the sun's rays strike the earth perpendicularly.

True north. The direction from any place on earth to the north pole.

UTM. Universal Transverse Mercator. Refers to the projection and the grid system based on it.

Value-by-area cartogram. A cartogram on which the size of the different units is proportional to some variable, such as population or income, not to actual geographic size.

Verbal scale. A statement in words of the map scale, such as 1 inch to the mile, or 1 centimeter to the kilometer.

Vertical exaggeration. On a profile, the "stretching" of the elevation with respect to the horizontal dimension. This is done to bring out small variations in relief and is needed and used more in relatively flat areas than in steep areas.

Vertical photograph. Photograph taken with the camera's axis perpendicular to the ground.

VFR chart. An aeronautical chart used by pilots flying under visual flight rules.

Virtual map. Basically, a map viewed on a computer screen.

Visible spectrum. That part of the electromagnetic spectrum that is visible to the human eye. It extends from violet at the short-wavelength end through blue, green, yellow, and orange to red at the long-wavelength end.

Visualization. Visualization involves exploration of data and seeing it in different ways. It is often associated with dynamic visual displays. The goal is to gain insight into the data. Some animated maps, such as flythroughs and flybys, are also called visualizations.

Wave length. The length from crest to crest of a wave in the electromagnetic spectrum. The longer the wave length, the lower the frequency.

Weather map. A special-purpose map used by meteorologists and others to show weather conditions.

Zenithal projection. Also called an azimuthal projection.

Zone of least deformation. The area on a projection that is most accurate. Sometimes called the area or zone of best representation.

APPENDIX B

Abbreviations

AAA	American Automobile Association
AAG	Association of American Geographers
AAHTSO	American Association of State Highway Transportation Officials
ASPRS	American Society of Photogrammetry and Remote Sensing
CAGIS	Cartography and Geographic Information Science
GIS	Geographic information system, geographic information science
GMT	Greenwich mean time
GPS	Global positioning system
IFR	Instrument flight rules
MEF	Maximum elevation figure
NACIS	North American Cartographic Information Society
NACO	National Aeronautical Charting Office
NASA	National Astronomical and Space Administration
NGS	National Geospatial-Intelligence Agency
NIMA	National Imagery and Mapping Agency
NOAA	National Oceanic and Atmospheric Administration
NPS	National Park Service
PLSS	Public Land Survey System
POI	Point of interest

RADAR   Radio Detection and Ranging

SPCS   Stored Program Control System

SYMAP   SYnagraphic MAPping

TIROS   Television Infrared Observational Satellite

TVA   Tennessee Valley Authority

UAV   Unmanned aerial vehicle

USDA   U.S. Department of Agriculture

USGS   U.S. Geological Survey

UTM   Universal transverse Mercator

VFR   Visual flight rules

APPENDIX C

Useful Statistics

EARTH DIMENSIONS

Equatorial diameter	7,926.68 miles; 12,756.4 kilometers
Polar diameter	7,899.99 miles; 12,713.45 kilometers
Equatorial circumference	24,902 miles; 40,074.89 kilometers
Meridional circumference	24,860 miles; 40,007.2 kilometers

DISTANCE MEASURES

English Units

1 rod	= 16.5 feet
	= 5.5 yards
1 furlong	= 660 feet
	= 220 yards
	= 40 rods
	= 1/8 statute mile
1 statute mile	= 63,360 inches
	= 5,280 feet
	= 1,760 yards

	= 320 rods
	= 8 furlongs
1 nautical mile	= 1 minute of arc along the equator
	= 6,076.1 feet
1 nautical mile	= 1.15 statute miles

Metric Units

1 millimeter	= 0.001 meter
1 centimeter	= 0.01 meter
1 meter	= 1,000 mm
	= 100 cm
1 kilometer	= 1,000 meters

METRIC/ENGLISH CONVERSIONS

1 millimeter = 0.039 inch

1 centimeter = 0.393 inch

1 meter = 39.37 inches

1 kilometer = 3,280.83 feet

1 kilometer = 0.621 statute miles

1 inch = 2.54 centimeters

1 foot = 30.48 centimeters

1 yard = 91.44 centimeters

1 yard = 0.914 meter

1 mile = 1.603 kilometers

PUBLIC LAND SURVEY SYSTEM

1 township is 6 miles × 6 miles

1 township contains 36 sections

1 section = 1 square mile = 640 acres

1 quarter section = 160 acres

1 quarter-quarter = 40 acres

1 acre   = 208'9" on a side

                  = 43,560 square feet

MAP SCALES

1:2,500	1 inch represents 0.0394 mile	1 cm rep 0.063 km
1:10,000	1 in represents 0.1578 mile	1 cm rep 0.1 km
1:20,000	1 inch represents 0.3156 mile	1 cm represents 0.2 km
1:24,000	1 inch represents 2,000 feet	1 cm represents 0.24 km
1:25,000	1 inch represents 0.3945 mile	1 cm represents 0.25 km
1:62,500	1 inch represents 0.986 mile	1 cm represents 0.625 km
1:63,360	1 inch represents 1 mile	1 cm represents 0.633 km
1:100,000	1 inch represents 1.578 miles	1 cm represents 1 km
1:500,000	1 inch represents 7.891 miles	1 cm represents 5 km
1:1,000,000	1 inch represents 15.78 miles	1 cm represents 10 km

APPENDIX D

Resources

CARTOGRAPHIC ORGANIZATIONS AND JOURNALS

Association of American Geographers, Cartography Specialty Group
www.csun.edu~hfgeg003/csg

British Cartographic Society
www.cartography.org.uk
Journal: *The Cartographic Journal*

Canadian Cartographic Association
www.cca-acc.org
Journal: *Cartographica*

Cartography and Geographic Information Society
www.cartogis.org
Journal: *CaGis Journal*

International Cartographic Association (ICA)
cartography.tuwein.ac.at/ica

North American Cartographic Information Society (NACIS)
www.nacis.org
Journal: *Cartographic Perspectives*

Society of Cartographers
www.soc.org.uk
Journal: *The Bulletin*

REGIONAL MAP SOCIETIES

California Map Society
   www.californiamapsociety.org/index.html

Chicago Map Society
   www.newberry.org/smith/cms/cms.html

New York Map Society
   www.nymapsociety.org

Texas Map Society
   libraries.uta.edu/txmapsociety

Washington Map Society
   home.earthlink.net/~docktor/washmap.htm
   Journal: *The Portolan*

HISTORY OF CARTOGRAPHY

The collections listed are but a few of the historic map collections, but all have maps viewable online.

British Library Map Collection
   www.bl.uk/onlinegallery/onlineex/mapsviews

David Rumsey (historic map collection online)
   www.davidrumsey.com

Herman Dunlop Smith Center for the History of Cartography
   www.newberry.org/smith/smithhome.html
   Newsletter: *Mapline*

History of Cartography Project
   www.press.uchicago.edu/books/HOC/index.html
   Volumes 1–3 of the *History of Cartography*
   Imago Mundi, The International Journal for the History of Cartography

Library of Congress
   www.loc.gov/maps

Newberry Library Map Collection
   www.newberry.org/maps-travel-and-exploration

Osher Map Library
   www.oshermaps.org

Stanford University Library Map Collection
   http://library.stanford.edu/subjects/rare-maps

U.S. GOVERNMENT AGENCIES

Bureau of the Census
   www.census.gov

National Aeronautics and Space Administration (NASA)
   www.nasa.gov

National Atlas
   www.nationalatlas.gov

National Geospatial-Intelligence Agency (NGA)
   https://www1.nga.mil

National Oceanic and Atmospheric Administration (NOAA)
   www.noaa.gov

U.S. Geological Survey (USGS)
   www.usgs.gov/pubprod

GOVERNMENT GEOGRAPHIC AND MAPPING AGENCIES

Because URLs are subject to change, these are not provided with the agency name. Looking up the name on a search engine will return the current Internet address. Also bear in mind that some countries consider maps, especially topographic maps, of strategic importance, and so they are not available to the public.

Australia	Geoscience Australia
Austria	Federal Office of Metrology and Surveying
Belgium	Belgian National Geographic Institute
Brazil	Brazilian Institute of Geography and Statistics
Canada	Centre for Topographic Information
Chile	Instituto Geográfica Militar
China	National Bureau of Surveying and Mapping
Colombia	Instituto Geográfico Agustin Codazzi
Czech Republic	Czech Office for Surveying, Mapping, and Cadastre
Denmark	Danish Geodata Agency

Estonia	Estonian Land Board
Finland	National Land Survey of Finland
France	Institut Géographique National
Germany	Bundesampt für Kartographie und Geodäsie
Great Britain	Ordnance Survey
Greece	Hellenic Military Geographical Service
Guatemala	Instituto Geográfico Nacional
Iceland	National Landsurvey of Iceland
India	Survey of India
Indonesia	Badan Informasi Geospasial
Iran	National Cartographic Center
Ireland	Ordnance Survey Ireland
Israel	Survey of Israel
Italy	Instituto Geografico Militare
Japan	Geospatial Information Authority of Japan
Mexico	Instituto Nacional de Estadistica y Geografia (INEGI)
Netherlands	Kadaster
New Zealand	Land Information, New Zealand
Nicaragua	INETER
Northern Ireland	Land and Property Services Northern Ireland
Norway	Statens Kartverk
Pakistan	Survey of Pakistan
Panama	Instituto Geográfica National
Portugal	Instituto Geográfico Português
Russia	Rostreestr
Slovenia	Surveying and Mapping Authority of the Republic of Slovenia
South Africa	Chief Directorate: National Geospatial Information
Spain	Instituto Geogràfico Nacional
Sri Lanka	Department of Survey
Sweden	Lantmäteriet
Switzerland	Swisstopo
Turkey	General Command of Mapping

United States   U.S. Geological Survey
Uruguay   Servicio Geografico Militar

PRIVATE CARTOGRAPHIC COMPANIES

Adventure Cycling
    Cycling maps
    adventurecycling.com

Benchmark Maps
    Atlas maps, phone apps
    www.benchmarkmaps.com

Delorme Maps and GPS
    www.delorme.com

Hammond Maps
    www.hammondmap.com

National Geographic Maps
    maps.nationalgeographiccom/maps

Rand McNally
    www.randmcnally.com

Tom Harrison Maps
    www.tomharrisonmaps.com

For a list of small independent mapping firms go to *www.nacis.org/index.cfm?x=16.*

APPENDIX E

Commonly Used Projections

MERCATOR

Classification	Cylindrical, conformal
Graticule	*Meridians:* equally spaced, straight, parallel lines. *Parallels:* unequally spaced, straight, parallel lines; spacing increases toward the poles.
Scale	True along the equator; constant along any parallel, increasing with distance from the equator; infinity at the poles.
Distortion	Increases away from the equator; areal distortion great in high latitudes.
Usage	Designed and recommended for navigation. Recommended and used for conformal maps of equatorial regions.
Other	All rhumb lines are straight lines.

TRANSVERSE MERCATOR

Classification	Cylindrical, conformal
Graticule	Spacing increases away from the central meridian. Equator is straight. *Parallels* are complex curves concave toward the nearest pole. Central *meridian* is straight, as are each meridian 90° from it; other meridians are complex curves concave toward the central meridian.
Scale	True only along the central meridian and minimal distortion within 15° of it.

Distortion	Increases rapidly away from the zone of least deformation.
Usage	Used by USGS for quadrangle maps at scales from 1:24,000 to 1:250,000. Such maps can be joined at their edges.
Other	Normally not used to show entire world.

MILLER CYLINDRICAL

Classification	Cylindrical, neither equal-area nor conformal.
Graticule	*Meridians:* equally spaced, straight, parallel lines. *Parallels:* unequally spaced, straight, parallel lines; closest at the equator.
Scale	True along the equator; constant along any parallel, changes with latitude and direction.
Distortion	Increases away from the equator.
Usage	World maps.

SINUSOIDAL

Classification	Pseudocylindrical; equal-area.
Graticule	*Meridians:* Central meridian is a straight line one-half as long as the equator; others are equally spaced sine curves. *Parallels:* equally spaced, straight, parallel lines; perpendicular to the central meridian.
Scale	True on the parallels and central meridian.
Usage	Atlas maps of South America and Africa; occasionally for world maps.
Other	Also called Sanson–Flamsteed projection.

MOLLWEIDE

Classification	Pseudocylindrical; equal-area.
Graticule	*Meridians:* The central meridian is a straight line one-half as long as the equator. Meridians 90° east and west of the central meridian form a circle; others form semiellipses. *Parallels:* unequally spaced, straight, parallel lines, perpendicular to the central meridian.
Scale	True along 40°N and S; constant along any given latitude.
Distortion	Severe near the outer meridians at high latitudes.
Usage	World atlas maps, especially thematic maps.
Other	Also called the homolographic; often used in interrupted form.

ECKERT IV

Classification	Pseudocylindrical; equal-area.
Graticule	*Meridians:* The central meridian is a straight line one-half the length of the equator; others are equally spaced semiellipses. *Parallels:* unequally spaced, straight, parallel lines; perpendicular to the central meridian; widest spacing is near the equator. Poles are lines one-half the length of the equator.
Scale	True along 40°30'; constant along any given latitude.
Distortion	Distortion free at 40°30'N and S at central meridian.
Usage	Thematic and other world maps in atlases and textbooks.

GOODE'S HOMOLOSINE

Classification	Pseudocylindrical; composite; equal-area.
Graticule	*Meridians:* in interrupted form, six straight central meridians; kink in meridians at 40°44'N and S. *Parallels:* straight parallel lines perpendicular to the central meridians. Poles are points.
Scale	True along every latitude between 40°44'N and S and along central meridian between 40°N and S.
Distortion	Same as sinusoidal between 40°44'N and S; same as Mollweide beyond this range.
Usage	Almost always presented in interrupted form; numerous world maps in atlases and textbooks.
Other	Composite of Mollweide (homolographic) and sinusoidal.

ROBINSON

Classification	Pseudocylindrical; neither equal-area nor conformal.
Graticule	*Meridians:* central meridian is a straight line; others resemble elliptical arcs. *Parallels:* straight, parallel lines; equally spaced between 38°N and S; spacing decreases beyond these limits.
Scale	True along 38°N and S; constant along any given latitude.
Distortion	Low within 45° of the central meridian and along the equator; no point is free of distortion.
Usage	Thematic world maps.

LAMBERT CONFORMAL CONIC

Classification	Conic; conformal.
Graticule	*Meridians:* equally spaced, straight lines converging at a common point (one of the poles). *Parallels:* unequally spaced, concentric circle arcs centered on the pole of convergence.
Scale	True along one or two chosen standard parallels; constant along any given parallel.
Distortion	Free of distance distortion along the one or two standard parallels.
Usage	Midlatitude regions of E–W extent.
Other	Cannot show entire earth.

ALBERS EQUAL-AREA CONIC

Classification	Conic; equal-area.
Graticule	*Meridians:* equally spaced, straight lines converging at a common point, normally beyond the pole. *Parallels:* unequally spaced concentric arcs. Poles are circular arcs.
Scale	True along the one or two standard parallels; constant along any given parallel.
Distortion	Free of angular and scale distortion only along the one or two standard parallels.
Usage	Frequently used for U.S. maps, thematic maps; recommended for equal-area maps of E–W midlatitude regions.

POLYCONIC

Classification	Polyconic; neither equal-area nor conformal.
Graticule	*Meridians:* The central meridian is a straight line; all others are complex curves. *Parallels:* equator is a straight line; all others are nonconcentric circular arcs spaced at true distances along the central meridian.
Scale	True along the central meridian and each parallel.
Distortion	Distortion-free only along the central meridian; extensive distortion if there is a great E–W range.
Usage	Sole projection for U.S. Geological Survey topographic maps until the 1950s.
Other	Not recommended for regional maps.

GNOMONIC

Classification	Azimuthal; perspective; neither equal-area nor conformal.
Graticule	Polar aspect. *Meridians:* equally spaced, straight lines intersection at the central pole. *Parallels:* unequally spaced, concentric circles centered on the pole; spacing increases toward the equator. Equator and opposite pole cannot be shown.
Scale	True only at the center.
Distortion	Only the center is distortion free; distortion increases rapidly away from the center.
Usage	To show great circle paths as straight lines; for navigation.
Other	All great circle arcs show as straight lines; can be used in polar, equatorial, and oblique aspects.

STEREOGRAPHIC

Classification	Azimuthal; conformal; perspective.
Graticule	Polar aspect. *Meridians:* equally spaced, straight lines intersecting at the central pole. *Parallels:* unequally spaced, concentric circles centered on the pole. Opposite pole cannot be shown.
Scale	True only at center.
Distortion	Only the center is distortion free.
Usage	For topographic maps of polar regions.
Other	All great or small circles show as circles or straight lines. Can be used in polar, equatorial, or oblique aspect.

ORTHOGRAPHIC

Classification	Azimuthal; perspective; neither conformal nor equal-area.
Graticule	Polar aspect. *Meridians:* equally spaced, straight lines intersecting at the central pole. *Parallels:* unequally spaced circles centered at the pole; spacing decreases away from the central pole. Only one hemisphere can be shown.
Scale	True only at the center.
Distortion	Only the center is distortion free.
Usage	Pictorial views of the earth and moon.

Other	Has the look of a globe. Can be used in polar, equatorial, or oblique aspect.

AZIMUTHAL EQUIDISTANT

Classification	Azimuthal; equidistant.
Graticule	Polar aspect. *Meridians:* equally spaced, straight lines intersecting at the central pole. *Parallels:* equally spaced circles centered at the pole. The entire earth can be shown.
Scale	True along any straight line radiating from the center.
Distortion	Only the center is distortion free.
Usage	Polar aspect is used for maps of the polar region; oblique aspect is frequently centered on major cities to show distances.
Other	Also called zenithal equidistant. Can be used in polar, equatorial, or oblique aspects.

AZIMUTHAL EQUAL-AREA

Classification	Azimuthal; equal-area.
Graticule	Polar aspect. *Meridians:* equally spaced straight lines intersecting at the central pole. *Parallels:* unequally spaced circles; spacing decreases away from the pole. Entire earth can be shown.
Scale	True only at the center.
Distortion	Only the center is distortion free.
Usage	Polar aspect is frequently used for atlas maps of polar regions.
Other	Can be used in polar, equatorial, or oblique aspects.

Bibliography

Aberley, Doug (Ed.). (1993). *Boundaries of Home: Mapping for Local Empowerment*. Philadelphia: New Society.

Abrams, Janet, and Peter Hall (Eds.). (2006). *Elsewhere: Mapping New Cartographies of Networks and Territories*. Minneapolis: University of Minnesota Design Institute.

Akerman, James R. (Ed.). (2006). *Cartographies of Travel and Navigation*. Chicago: University of Chicago Press.

American Society of Photogrammetry. (1960). *Manual of Photographic Interpretation*. Washington, DC: Author.

Andrews, J. H. (1999). "What Was a Map? The Lexicographer's Reply." *Cartographica: The International Journal for Geographic Information and Geovisualization*, 33(4), 1–12.

Andrews, J. H. (2009). *Maps in Those Days: Cartographic Methods before 1850*. Dublin: Four Courts Press.

Bagrow, Leo, and Raleigh Skelton. (1966). *The History of Cartography*. Cambridge, MA: Harvard University Press.

Brown, Lloyd. (1949). *The Story of Maps*. New York: Bonanza Books.

Bugayevskiy, Lev M., and Snyder, John P. (1995). *Map Projections: A Reference Manual*. London: Taylor & Francis.

Calder, Nigel. (2003). *How to Read a Nautical Chart*. Camden, ME: International Marine/McGraw-Hill.

Campbell, James B., and Randolph H. Wynne. (2011). *Introduction to Remote Sensing* (5th ed.). New York: Guilford Press.

Campbell, John. (2000). *Map Use and Analysis*. New York: McGraw-Hill.

Cartwright, William, Michael P. Peterson, and Georg Gartner. (2007). *Multimedia Cartography* (2nd ed.). New York: Springer.

Crampton, Jeremy W. (2010). *Mapping: A Critical Introduction to Cartography and GIS*. Chichester, UK: Wiley-Blackwell.

Deetz, Charles H., and Oscar S. Adams. (1945). *Elements of Map Projection: With Applications to Map and Chart Construction* (Special Publication No. 68, 5th ed. rev.). Washington, DC: U.S. Government Printing Office. (Reprint edition available from Nabu Press, 2011).

Dodge, Martin, Rob Kitchin, and Chris Perkins. (2011). *The Map Reader: Theories of Mapping Practice and Cartographic Representation*. West Sussex, UK: Wiley-Blackwell.

Dorling, Daniel, and David Fairbairn. (1997). *Mapping: Ways of Representing the World*. Essex, UK: Prentice-Hall.

Drury, George H. (1952). *Map Interpretation* (2nd ed.). London: Pitman.

Evans, R. T., and H. M. Frye. (2009). *History of the Topographic Branch (Division)* (U.S. Geological Survey Circular 1341). Available at *http://pubs.usgs.gov/circ/1341*.

Federal Aviation Administration. (2012). *The Aeronautical Chart Users Guide* (9th ed.). New York: Skyhorse.

Fonstad, Karen Wynn. (2001). *The Atlas of Middle Earth* (rev. ed.). New York: Mariner Books.

Garfield, Simon. (2012). *Off the Map: A Mind Expanding Exploration of the Way the World Looks*. New York: Gotham Books.

Gibson, Rich, and Schuuler Erie. (2006). *Google Maps Hacks*. Sebastapol, CA: O'Reilly Media.

Greenhood, David. (1944). *Down to Earth, Mapping for Everybody*. New York: Holiday Books.

Greenhood, David. (1964). *Mapping*. Chicago: University of Chicago Press.

Harris, Lucia Carolyn. (1960). *Sun, Earth, Time, and Man*. Chicago: Rand McNally.

Hoogvliet, Margriet. (1996). "The Mystery of the Makers; Did Nuns Make the Ebsdorf Map?" *Mercator's World*, 1(6), 16–21.

Hubbard, Bill, Jr. (2009). *American Boundaries: The Nation, the States, the Rectangular Survey*. Chicago: University of Chicago Press.

Huth, John Edward. (2013). *The Lost Art of Finding Our Way*. Cambridge, MA: Belknap Press of Harvard University.

Jacobs, Ryan. (2013). "A Lost Scottish Island, George Orwell, and the Future of Maps." *The Atlantic*. Retrieved September 23, 2013, from *www.theatlantic.com/international/archive/2013/07*.

Jennings, Ken. (2011). *Maphead*. New York: Scribner.

Jensen, John R. (2006). *Remote Sensing of the Environment* (2nd ed.). Englewood Cliffs, NJ: Prentice-Hall.

Johnson, Hildegard Binder. (1976). *Order upon the Land*. New York: Oxford University Press.

Johnston, Andrew K. (2004). *Earth from Space* (Smithsonian National Air and Space Museum). Buffalo, NY: Firefly Books.

Kain, Roger J. P., and Elizabeth Baigent. (1992). *The Cadastral Map in the Service of the State: A History of Property Mapping*. Chicago: University of Chicago Press.

Kanas, Nick. (2007). *Star Maps: History, Artistry, and Cartography*. Chichester, UK: Springer Praxis.

Keates, J. S. (1982). *Understanding Maps*. New York: Halsted Press.

Kennedy, Melita, and Steve Kopp. (2000). *Understanding Map Projections*. Redlands, CA: ESRI Press.

Kimmerling, Jon, Aileen Buckley, Phillip Muehrcke, and Juliana Muehrcke. (2010). *Map Use: Reading, Analysis and Interpretation* (7th ed.). Redlands, CA: ESRI Press.

Kjellstrom, Björn. (2009). *Be Expert with Map and Compass* (3rd ed.). Hoboken, NJ: Wiley.

Klinghoffer, Arthur Jay. (2006). *The Power of Projections: How Maps Reflect Global Politics and History*. Westport, CT: Praeger.

Kraak, Menno-Jan. (2011). "Is There a Need for Neo-Cartography?" *Cartography and Geographic Information Science, 38*(2), 73–78.

Kraak, Menno-Jan, and Ferjan Ormeling. (2010). *Cartography: Visualization of Spatial Data* (3rd ed.). New York: Guilford Press.

Letham, Lawrence. (2008). *GPS Made Easy* (5th ed.). Seattle, WA: The Mountaineers.

Lewis, Tom. (1997). *Divided Highways*. New York: Penguin Books.

Lillesand, Thomas, Ralph W. Kiefer, and Johnathan Chapman. (2007). *Remote Sensing and Image Interpretation* (6th ed.). Hoboken, NJ: Wiley.

Linklater, Andro. (2002). *Measuring America: How an Untamed Wilderness Shaped the United States and Fulfilled the Promise of Democracy*. New York: Walker.

Lobeck, Armin K. (1956). *Things Maps Don't Tell Us*. New York: Macmillan.

MacEachren, Alan. (1994). *Some Truth with Maps: A Primer on Symbolization and Design*. Washington, DC: Association of American Geographers.

MacEachren, Alan. (1994, 2004). *How Maps Work: Representation, Visualization, and Design*. New York: Guilford Press.

Mitchell, Tyler. (2005). *Web Mapping Illustrated*. Sebastapol, CA: O'Reilly Media.

Monmonier, Mark. (1985). *Technological Transition in Cartography*. Madison: University of Wisconsin Press.

Monmonier, Mark. (1993). *Mapping It Out: Expository Cartography for the Humanities and Social Sciences*. Chicago: University of Chicago Press.

Monmonier, Mark. (1995). *Drawing the Line*. New York: Henry Holt.

Monmonier, Mark. (1996). *How to Lie with Maps* (2nd ed.). Chicago: University of Chicago Press.

Monmonier, Mark. (1999). *Air Apparent: How Meteorologists Learned to Map, Predict, and Dramatize Weather*. Chicago: University of Chicago Press.

Monmonier, Mark. (2004). *Rhumb Lines and Map Wars: A Social History of the Mercator Projection*. Chicago: University of Chicago Press.

Monmonier, Mark, and George Schnell. (1988). *Map Appreciation*. Englewood Cliffs, NJ: Prentice Hall.

Morton, Oliver. (2002). *Mapping Mars: Science, Imagination, and the Birth of a World*. New York: Picador USA.

Ovenden, Mark. (2007). *Transit Maps of the World*. New York: Penguin.

Paine, David P., and James D. Kiser. (2012). *Aerial Photography and Image Interpretation* (3rd ed.). Hoboken, NJ: Wiley.

Parker, Mike. (2009). *Map Addict*. London: Collins.

Pearce, Margaret, and Owen Dwyer. (2010). *Exploring Human Geography with Maps* (2nd ed.). New York: W. H. Freeman.

Peterson, Michael P. (1995). *Interactive and Animated Cartography*. Englewood Cliffs, NJ: Prentice-Hall.

Post, J. B. (1979). *An Atlas of Fantasy* (rev. ed.). New York: Ballantine Books.

Price, Edward T. (1995). *Dividing the Land: Early American Beginnings of Our Private Property Mosaic*. Chicago: University of Chicago Press.

Reeves, Robert G. (Ed. in Chief). (1975). *Manual of Remote Sensing: Volume I. Theory, Instruments, and Techniques*. Falls Church, VA: American Society of Photogrammetry.

Reeves, Robert G. (Ed. in Chief). (1975). *Manual of Remote Sensing: Volume II. Interpretation and Applications*. Falls Church, VA: American Society of Remote Sensing.

Rhind, D. W., and D. R. F. Taylor. (Eds.). (1989). *Cartography Past, Present and Future*. London: Elsevier.

Robinson, Arthur H. (1952). *The Look of Maps: An Examination of Cartographic Design*, reprint edition 2010. Redlands, CA: ESRI Press.

Robinson, Arthur H. (1982). *Early Thematic Mapping in the History of Cartography*. Chicago: University of Chicago Press.

Robinson, Arthur H., and Barbara Bartz Petchenik. (1976). *The Nature of Maps: Essays toward Understanding Maps and Mapping*. Chicago: University of Chicago Press.

Snyder, John P. (1993). *Flattening the Earth: Two Thousand Years of Map Projections*. Chicago: University of Chicago Press.

Sobel, Dava. (1995). *Longitude: The True Story of a Lone Genius Who Solved the Greatest Scientific Problem of His Time*. New York: Walker.

Taylor, D. R. Fraser. (2005). *Cybercartography: Theory and Practice*. Amsterdam: Elsevier.

Taylor, D. R. Fraser, and Stephanie Pyne. (2010). "The history and development of the theory and practice of cybercartography." *International Journal of Digital Earth*, 3(1), 2–15.

Taylor, E. G. R. (1971). *The Haven-Finding Art: A History of Navigation from Odysseus to Captain Cook* (augmented ed.). New York: American Elsevier.

Thompson, Morris M. (1988). *Maps for America* (3rd ed.). Washington, DC: USGS.

Thrower, Norman J. W. (1966). *Original Survey and Land Subdivision: A Comparative Study of the Form and Effect of Contrasting Cadastral Surveys*. Chicago: Association of American Geographers and Rand McNally.

Thrower, Norman J. W. (2008). *Maps and Civilization* (3rd ed.). Chicago: University of Chicago Press.

Tucker, Patrick. (2013, July–August). "Mapping the Future with Big Data." *The Futurist*, 46(4). Retrieved from *www.wfs.org/futurist/2013-issues-futurist/july-august-2013-vol-47-no-4*.

Tyner, Judith. (1969). "Early Lunar Cartography." *Surveying and Mapping*, 29(4), 583–596.

Tyner, Judith. (2010). *Principles of Map Design*. New York: Guilford Press.

U.S. Department of the Army. (2011). *Map Reading and Land Navigation*. Washington, DC: Author.

Wade, Tasha, and Shelly Sommer. (Eds.). (2006). *A to Z GIS*. Redlands, CA: ESRI Press.

White, C. Albert. (1991). *History of the Rectangular Survey System*. Washington, DC: U.S. Government Printing Office.

Whittaker, Ewen A. (1999). *Mapping and Naming the Moon*. Cambridge, UK: Cambridge University Press, 1999.

Wilford, John Noble. (2000). *The Mapmakers* (2nd ed.). New York: Vintage Books.

Winchester, Simon. (2001). *The Map That Changed the World: William Smith and the Birth of Modern Geology*. New York: HarperCollins.

Wood, Denis. (1992). *The Power of Maps*. New York: Guilford Press.

Wood, Denis. (2010). *Rethinking the Power of Maps*. New York: Guilford Press.

Wright, John K. (1942). "Mapmakers Are Human: Comments on the Subjective in Maps." *The Geographical Review*, 23(4), 527–544. [Reprinted in Dodge, Martin, Rob Kitchin, and Chris Perkins (Eds.). (2011). *The Map Reader: Theories of Mapping Practice and Cartographic Representation*. West Sussex, UK: Wiley-Blackwell.]

The History of Cartography, Volumes 1–6. (Volumes 1, 2, and 3 are viewable online.) Chicago: University of Chicago Press.

   Volume 1. (1987). *Cartography in Prehistoric, Ancient and Medieval Europe and the Mediterranean.*

   Volume 2, Book 1. (1992). *Cartography in the Traditional Islamic and South Asian Societies.*

   Volume 2, Book 2. (1995). *Cartography in the Traditional East and Southeast Asian Societies.*

   Volume 2, Book 3. (1998). *Cartography in the Traditional African, American, Arctic, Australian, and Pacific Societies.*

   Volume 3. *Cartography in the European Renaissance.*

   Volume 4. *Cartography in the European Enlightenment* (forthcoming).

   Volume 5. *Cartography in the Nineteenth Century* (forthcoming).

   Volume 6. *Cartography in the Twentieth Century* (forthcoming).

Index

Entries in Appendix A: Glossary and Appendix E: Commonly Used Projections are **Bold Face**; *f*, figure/illustration; *t*, table; *p*, plate

About the Author

Judith A. Tyner, PhD, is Professor Emerita of Geography at California State University, Long Beach. She taught in the Geography Department for over 35 years, where she served as Department Chair for 6 years and as Director of the Cartography/GIS Certificate Program from its inception (1980) until her retirement. While at California State Dr. Tyner taught beginning and advanced cartography, map reading and interpretation, history of cartography, and remote sensing. She is a member of the Association of American Geographers, the North American Cartographic Information Society, the Cartography and Geographic Information Society, and the California Map Society. She is the author of several textbooks, including *Principles of Map Design*, and over 30 articles.